Safety Management Systems

Safety Management Systems

Applications for the Aviation Industry

Edited by
Mark A. Friend, Alan J. Stolzer,
and Marisa D. Aguiar

Bernan Press

Lanham • Boulder • New York • London

Published by Bernan Press
An imprint of The Rowman & Littlefield Publishing Group, Inc.
4501 Forbes Boulevard, Suite 200, Lanham, Maryland 20706
www.rowman.com

800-865-3457; info@bernan.com

6 Tinworth Street, London SE11 5AL, United Kingdom

Paperback ISBN: 978-1-64143-361-7
E-ISBN: 978-1-64143-362-4

Contents

List of Figures

List of Tables

List of Abbreviations

AC	Advisory Circular
AHP	Analytical Hierarchy Process
AIHA	American Industrial Hygiene Association
ANSI	American National Standards Institute
ASAP	Aviation Safety Action Program
ASP	Air Service Provider
ATM	Air Traffic Management
AVS	Aviation Safety
CASA	Civil Aviation Safety Authority
CCOHS	Canadian Centre for Occupational Health and Safety
CEO	Chief Executive Office
CFR	Code of Federal Regulations
COO	Chief Operating Officer
CRM	Crew Resource Management
CTD	Cumulative Trauma Disorder
CWT	Center Wing Fuel Tank
DAHP	Delphi Analytical Hierarchy Process
EASA	European Aviation Safety Agency
EMP	Emergency Management Plan
ERP	Emergency Response Plan
FAA	Federal Aviation Administration
FBO	Fixed-Based Operator
FOD	Foreign Object Debris
FOQA	Flight Operations Quality Assurance
GAO	Government Accounting Office
HFACS	Human Factors Analysis and Classification System
HSE	Health and Safety Executive

ICAO	International Civil Aviation Organization
IEP	Internal Evaluation Program
ILO	International Labor Organization
ISO	International Organization for Standardization
IT	Information Technology
LOSA	Line Operations Safety Assessment
MOU	Memorandum of Understanding
MSD	Musculoskeletal Disorder
NASA	National Aeronautics and Space Administration
NSTB	National Transportation Safety Board
OSHA	Occupational Safety and Health Administration
OHSAS	Standard Occupational Health and Safety Assessment Specification
OHSMS	Occupational Health and Safety Management System
PDPC	Process Decision Program Chart
PRA	Probabilistic Risk Assessment
QMS	Quality Management System
RAA	Roles, Responsibility, and Authority
ROI	Return on Investment
SA	Safety Assurance
SARP	Standards and Recommended Practices
SHELL	Software-Hardware-Environment-Liveware-Liveware
SME	Subject Matter Expert
SM ICG	Safety Management International Collaboration Group
SMS	Safety Management Systems
SOP	Standard Operating Procedures
SPI	Safety Performance Indicator
SRM	Safety Risk Management
SSP	State Safety Program
SWOT	Strength-Weakness-Opportunity-Threat
TEM	Threat and Error Management
TIC	Total Incident Cost
TQM	Total Quality Management
TWA	Trans World Airlines
VDRP	Voluntary Disclosure Reporting Program
VPP	Voluntary Protection Programs
WHO	World Health Organization

Preface

This book was developed in response to a need for a comprehensive overview on the utilization of Safety Management Systems, specifically tailored toward industrial and aviation populations. As a first edition, the book was written with the following objectives in mind:

- It should be easily comprehensible by those who lack experience or prior exposure to the concepts of Safety Management Systems in industrial and aviation environments.
- It should be on a writing level easily understood by a majority of the population, and yet be useful as a text for college courses.
- It should be comprehensive enough to cover the major topics of concern relevant to the design, implementation, and management of a Safety Management System in both industrial and aviation environments.

As the title implies, this book covers the foundational aspects of designing, implementing, and managing a Safety Management System, from ensuring regulatory compliance is maintained to guidance on how to handle accident/incident investigations. Unlike other books exploring only the general principles of Safety Management Systems from a theoretical perspective, this text attempts to balance the necessary, foundational knowledge with practical applications to guide the industrial and aviation professional in establishing, managing, and controlling an effective Safety Management System, tailored to the specific needs of the organization.

The authors wish to thank Embry-Riddle Aeronautical University's PhD in Aviation students, candidates, and graduates for their contributions to the development of this book.

Marisa D. Aguiar
PhD in Aviation Candidate
Doctoral Graduate Research Assistant
College of Aviation
Embry-Riddle Aeronautical University
Daytona Beach, Florida

Acknowledgments

Thank you to the students in the PhD in Aviation at Embry-Riddle Aeronautical University. They contribute; we all learn.

Mark A. Friend

I'd like to acknowledge the contributions of our excellent students in the PhD in aviation program. I am continually amazed by the professional accomplishments, the drive, and the keen intellect of our doctoral students.

Alan J. Stolzer

To Jamison, for all the sacrifices you make so I can continue to pursue my dreams. To my mom, for your incessant support and encouragement. You are both my wings. A special thanks to my fellow editors, Drs. Mark Friend and Alan Stolzer, for the opportunity to participate in the development of this text.

Marisa D. Aguiar

Introduction

Marisa D. Aguiar

IMPORTANCE OF SMS

Safety Management Systems (SMS) requires an organization to thoroughly examine its operations and the decisions driving those operations. SMS is an effectively designed hierarchy that, by its nature, has the potential to truly infiltrate all aspects of a corporation's policy and procedures, allowing for adaptions to changes within the industry while working with limited resources. SMS actively promotes the continuous improvement of safety, utilizing specific methods to analyze, assess, and control risks by predicting hazards from employee reports, audits, and continuous, data-collection processes. SMS assists organizations in complying with existing federal regulations while actively predicting the need for future action, by sharing knowledge and information both within the organization and industry. Finally, an effective SMS is one that will enhance the safety attitudes of the entire organization by promoting a healthy safety culture from the highest-level of management to those out on the shop floor. The changes that occur through weaving SMS into an organization are designed to utilize reactive, proactive, and predictive thinking to foster a comprehensive view of safety.

SMS has generated wide support in a variety of diverse industries including aviation, occupational health, construction, oil, highway, electrical, and others, as an effective approach to improving safety. In addition to refining the safety of operations, SMS has financial benefits to organizations through increasing efficiency and avoiding the costs associated with errors. SMS integrates safety concepts into repeatable, proactive processes within a single system, approaching safety management as a fundamental business process, rather than just another safety program. The structure of SMS provides organizations with greater insight into their operational environment. According to the Federal Aviation Administration, some organizations have found the financial benefits of implementing a SMS materialize early in the implementation process.

SMS is the product of a continuing evolution in aviation safety. Historically, the aviation industry has been exceedingly reactive in its approach to safety. Over time, careful regulation of aviation activities, operational experience, and improvements in technology have contributed to significant gains in safety. In the next major phase of improvement to safety, a focus on individual and crew performance further reduced accidents. Thus, the progression of proactive safety within aviation has been and continues to be an evolving process. Each contribution has led to significant gains. However, even with these significant advances, there are still opportunities to reduce the number of accidents, incidents, and errors. Careful analysis of accident and incident reports typically reveals multiple opportunities for actions that could have broken the chain of events leading up to an occurrence and possibly prevented the occurrence entirely. These opportunities represent the organization's role in accident prevention. SMS is an approach to improving safety at the organizational level.

One of the purposes of SMS is to improve the safety performance, and therefore reduce the exposure to risk of having an accident. It is not focused on the safety record per se. Quality systems are focused on continuous improvement also, but through improving the production record rate. This is another source of confusion between the two management system concepts; improving a safety record, is not the same as improving safety performance. There are many aviation companies that have extremely good safety records but are operating with risky behavior or inadequate organizational structures and have just not had an accident yet. A good safety record, just like a good quality record, does not guarantee safety.

Steven C. McNeely, Manager, Safety Management Systems, Jet Solutions, L.L.C, published by "Flight Safety Information February 12, 2010 No.034", Lessons Learned from Toyota—2010-QMS vs. SMS.

WHO IS RESPONSIBLE FOR SAFETY WITHIN A SMS

As described by Stolzer, Halford, and Goglia (2008), *the responsibility for safety is pushed out onto the shop floor*. Within the context of a healthy safety culture, employees of an organization with a mature SMS understand they are fully accountable for their actions and play an active role in reporting, encouraging, and maintaining safety. However, this must be structured into the foundation of the organization's safety culture. The commitment of top management is vital to the overall success of the organization's SMS, specifically in regard to the health of the organization's safety culture. When the highest levels of management have strong commitments to safety, these values and attitudes trickle down to all levels of the organization.

COMMON MISCONCEPTIONS ADDRESSED

According to the Federal Aviation Administration, SMS applies system safety concepts and adds formal system safety management concepts, thereby creating an avenue for a more effective interface between product/service providers and their oversight organizations. SMS has become the next step in the evolution of system safety.

While larger organizations may have specific personnel dedicated to safety, such as a Director of Safety, safety and quality auditors and analysts, and investigators; it is not required. Rather, SMS is a set of management practices and can be implemented into organizations without hiring full-time safety personnel. Further, SMS is intended to be adaptable to fit the needs of the particular organization; therefore, resource allocation should be appropriate for the size of the organization. Finally, if an organization has a quality management system, it should not conflict with the SMS. Voluntary Employee Reporting programs are a requirement for SMS, but no specific programs are specified.

THE FOUR COMPONENTS OF SMS

Outlined in the FAA Advisory Circular 120-92, SMS is structured upon four components of safety management: safety policy, safety risk management, safety assurance, and safety promotion. The following provides a brief description of the four components, based on the descriptions within Advisory Circular 120-92:

Safety Policy. Safety policy establishes management's commitment to the continuous improvement of safety. This component encompasses the methods, processes, and organizational structure needed to meet the predetermined safety objectives of the organization, building upon the processes and procedures already in place. It is through the creation of safety policy that managerial commitment is established and cross-organizational communication and cooperation is facilitated.

Safety Risk Management. Within the safety risk management component of SMS, the relevant hazards are identified and risk controls are adapted, based on the assessment of acceptable risk. This is a formal process that consists of describing the system, identifying hazards, assessing risk, analyzing risk, and controlling risk.

Safety Assurance. Safety assurance evaluates the continued effectiveness of the implemented risk control strategies and supports the identification of new hazards via a feedback system, such as audits, safety surveys, and employee-reporting methods. The new information discovered during the feedback process is then utilized to ensure continuous improvement within the organization.

Safety Promotion. Safety promotion includes training, communication, and other actions required to create and maintain a positive safety culture within all levels of the organization. Activities promoting safety within the SMS framework include providing SMS training, advocating for a positive safety culture, promoting safety communication and awareness, and disseminating safety lessons learned. Effective safety promotion requires commitment from every employee and manager within the organization.

SCALING APPROPRIATELY TO OPERATIONAL RISKS

A one-size-fits-all approach it not optimal for any SMS. Safety managers, if applicable, overseeing the SMS implementation process must take into account the specific operational risks that exist within the context of their own particular organization. Various factors will influence the operational risk, such as the organization's unique operating procedures and the size of the organization. However, the primary objective of SMS–safety–should be at the forefront, regardless of risk continuum or organizational size. Risk must be managed on a scale beyond any organization's operational specifications and boundaries. When attempting to scale the SMS to any organization's unique needs, it is key to ensure all stakeholders are active participants with common goals and shared accountability.

MAINTAINING A DYNAMIC RISK MANAGEMENT SYSTEM

Risk management is considered to be the core process of an effective SMS for identifying hazards and assessing risk within an organization. In large, dynamic, high-risk organizations, traditional safety assessment processes, such as risk matrices, hazard logs, and other conventional risk management methodologies, may be insufficient. Some organizations may find value in proactively identifying the current versus actual risk of their system, for the purpose of taking necessary mitigation actions early to avoid a serious accident or incident occurring. Regardless of the approach to risk management, the process must be scaled to the needs of the organization and subjected to continuous improvement and assessment.

The chapters within this book are designed to provide the aviation industry professional with the information required to implement and utilize SMS practices from the legal considerations required for implementation through providing guidance on navigating the accident investigation process. Finally, the text will conclude with a practical example of SMS implementation within the rotorcraft industry.

REFERENCES

Federal Aviation Administration. (2010). Lessons learned from Toyota, 2010. *Flight Safety Information No.034*. Retrieved from https://www.faa.gov/about/initiatives/sms/ explained/quality_and_safety_management/media/QMS_vs_SMS_Lessons_Learned_from_Toyota.pdf

Federal Aviation Administration. (2010). Safety management systems for aviation service providers. *Advisory Circular 120-92a*. Retrieved from https://www.faa.gov/document Library/media/Advisory_Circular/AC%20120-92A.pdf

Stolzer, A. J., Halford, C. D. & Goglia, J. J. (2008). *Safety management systems in aviation*. Burlington, VT: Ashgate Publishing Company.

Chapter 1

The Legal Aspects of a Safety Management System (SMS)

Cory J. Trunkhill, Kris Ostrowski, Gregory S. Woo, and David A. Carroll

Modern aviation safety management must have a foundation rooted in legal doctrine. Aviation organizations owe it to the general public to maintain a state of safe and secure operations. An essential part of this goal is a safe, secure, and healthful work environment for the employees of these aviation organizations. The laws and statues imposed on aviation by governmental regulatory committees are meant to ensure certain definable and achievable safety standards are met.

Legal doctrine is a component upon which safety management rests. Clear directives and administration of safety expectations within the aviation community need to assert the reasonable expectation of safe operations and environment to the public, consumers, and workers within the field. If the legal doctrine is ambiguous, unclear, or incomprehensible, subsequent safety regulation and continuity of administration will suffer. Critical to understanding the relationships between legal safety doctrine and administrative practices will be the Occupational Safety and Health Administration's (OSHA) effect and relationship to aviation entities to include the National Transportation and Safety Bureau (NTSB), Federal Aviation Administration (FAA), and the National Aeronautics and Space Administration (NASA). The resulting construct has defined how OSHA regulations, guidance, education, and training contribute to safety management systems (SMS).

OSHA REGULATIONS

OSHA, as a government agency specializing in worker safety and work conditions regardless of vocation, was established under the Department of Labor in 1970. The original legislative act empowering the agency to improve and regulate worker safety was amended in 2000 to further protect industry whistleblowers against retribution for reporting unsafe acts and company violations of federal code (OSHA, 2017a). As applied to the aviation industry, OSHA regulations continue to impact the safety and security of both passengers and industry workers. Engaging with the agency is essentially a trifecta, as any stakeholder or participant can review, report, and confer with agency representatives on regulations or safety concerns.

OSHA's mission is to assure safe and healthful working conditions for men and women by setting and enforcing standards and by providing training, outreach, education, and assistance (OSHA, 2017a). First, OSHA aims to assure a safe and healthful work environment by setting standards, regulations, and guidelines. These can be easily accessed via publications or online. Second, the agency enforces those standards by way of inspections, and consequences typically follow when standards are not being upheld. Consequences come in the form of fines, litigation, and even temporarily prohibiting the operation of the organization until compliancy is met. Lastly, OSHA provides training, outreach, education, and assistance (OSHA, 2017a). The agency has continued-education courseware available to the public at no charge. Additionally, government publishing provides hard copies of literature for workers and management to have on hand. Training courses and site visitation by agency representatives help support best practices and recommended procedures to ensure and improve safety conditions.

There are thousands of U.S. OSHA standards in existence, and any number may be applicable within various businesses, depending upon their specific circumstances. Employers are fully expected to follow OSHA guidelines. Violations can result in substantial penalties.

Workplace violations may either be reported by so called *whistleblowers* or come as a result of agency inspection violations (Schmidt, 2002). There is the opportunity, in formal hearings, for aviation operations to defend themselves and come to terms of agreeance. The terms may include adjudication of fines based on positive corrective measures under a timeline. There are other instances warranting more serious degrees of punishment based on gross maleficence (Schmidt, 2002). In these instances, data, evidence, and interviews are gathered for federal prosecutors to build a case against the organization. Nevertheless, the agency affords operators the opportunity to correct unsafe actions and workplace violations before drastic measures are taken, within reason.

OSHA will levy penalties against aviation operations for instances of gross negligence, non-compliance, and unsafe acts (Schmidt, 2002). These penalties are to ensure workers and the public remain safe and healthful within the business infrastructure. Fines, forfeitures, and cease-and-desist notices are meant as corrective actions. From formal notices to shuttering operations, OSHA employs a corrective approach to addressing violations of the Code of Federal Regulations (CFR). OSHA (U.S.) currently levies maximum penalty fines of up to $13,260 for each serious violation, and fines up to $132,598 for a repeated or willful violation of standards (OSHA, 2019). As mentioned previously, serious repercussions can befall a business with assessed violations, both before a mishap occurs and after. For instance, prior to the Deepwater Horizon/British Petroleum (BP) oil spill in 2010, OSHA inspected BP in 2005 and cited fourteen violations amounting to $21,361,500 in penalties. In 2009, OSHA re-inspected BP and found 848 violations and assessed $81,340,000 in penalties. Businesses can not only be held legally accountable for assessed violations but may also be held accountable for voluntary guidelines considered an industry standard.

Voluntary Safety Management

Companies are encouraged, through a merit-based system, to exhibit voluntary safety protection programs for their employees. These are defined through OSHA literature and online information. Currently, a tiered recognition system for excellence in voluntary safety management promotion exists to encourage companies to see the value in preventing compensation for worker injury or illness (OSHA, 2015). The significance of this approach is that the company will save time, effort, and funds by proactive safety management in achieving voluntary compliance and excellence in worker safety.

There are currently three recognized voluntary protection programs (VPP) afforded to the industry. The *Demonstration* designation recognizes employer safety training programs, which differ significantly from OSHA qualifications (OSHA, 2017b). The *Merit* basis is a qualifier that denotes while VPP is being practiced by the company; improvements in training and application need to be addressed. Lastly, the *Star* designation is the pinnacle of achievement through a VPP program. It signifies that the current training problem addresses OSHA recommendations concerning regular employee sessions that address safety and wellness (OSHA, 2017b).

The value in a voluntary safety management program is threefold. First, the employers manage risk and hazards at the lowest level with available resources. Workers not sick nor injured do not expend company resources. Second, workers gain self-awareness and corporate buy-in to develop a safety culture. Third, voluntary safety and wellness classes, as adopted by management, aid in the promotion of safety excellence. More information about these voluntary programs can be found via the official OSHA website.

FAA SAFETY AND HEALTH PROGRAM MANAGEMENT GUIDELINES

Some safety regulations impacting aviation are rooted in the Code of Federal Regulations (CFR). Specifically, Title 14 governs Federal Air Regulations (FAR) and the associated subsidiary Parts (FAA, 2017). These Parts provide governance on the safe operation of aircraft, conduct of crewmembers, and ground personnel. They also dictate the manner in which the public engages with air transportation relative to the part operation type (FAA, 2017). CFR 14 Part requirements are listed in in table 1.1. They specify safety and health measures distinct to inspection and enforcement, corrective actions, safety of passenger operations, and airworthiness standards. While table 1.1 is not all inclusive, these are but a sample of the dozens of regulations operators must abide by. However, those listed pertain to the basis of SMS formation.

The regulations serve as guidance to establishing the SMS. By way of mandatory minimum requirements and voluntary protection programs, safety cultures of excellence may develop in any size organization. Further, they also serve a role in enforcement, reporting, continuous education, and process improvement.

Leadership. The agency sponsors training and programs to encourage the industry to foster and maintain safety management practices at the highest levels. Open communication with workers and a proactive approach to health and safety are critical to developing a safety culture and overall workplace excellence (Manuele, 2017; OSHA, 2017a). By utilizing a top-down safety management approach, corporate buy-in by all stakeholders enables accountability, responsibility, and embodiment of genuine care for the health and welfare of all workers.

Hazard identification and assessment. Continuous review of known hazards, along with detection and documentation of emerging ones, ensure managers and personnel are always cognizant of them. Risk mitigations and avoidance can be implemented at every level; although, recognition and reporting are key to initiating a response to all hazards. Managers will, upon

Table 1.1 CFR 14 Part Regulations (FAA, 2017)

Part Number	*Description*
13	Investigation and Enforcement Procedures
39	Airworthiness Directives
121	Operating Requirements: Domestic, Flag, and Supplemental Operations
135	Operating Requirements: Commuter and On Demand Operations and Rules Governing Persons on Board Such Aircraft
141	Flight Schools

identification of a new hazard, initiate documentation to assess the hazard, which will determine how best to handle the situation (Manuele, 2017; OSHA, 2017a).

Hazard prevention and control. Closely associated with hazard identification and assessment, prevention and control seek to maintain a continuity of safety culture. The entirety of a workplace is encouraged to preempt, mitigate, or control identified hazards. This is accomplished through employee empowerment to make recommendations and follow-through for review. While a process must be established to accomplish this, thorough documentation and review by leadership is key to any selected course of action. The review process must be able to further evaluate the efficacy of the selected method for hazard control (Manuele, 2017; OSHA, 2017a).

Education and training. Minimum training requirements at every level of an organization need to be accessible and recorded. Accounting for proper instruction facets, based on skills and duties, must be reviewed and even demonstrated in certain instances (Manuele, 2017; OSHA, 2017a). Evaluation and inspection of mandated safety requirements ensure hazards are recognized and properly controlled. This reinforces, within the worker, the necessity for such education and training to help maintain the wellbeing of themselves and their fellow employees.

Program evaluation and improvement. Hazard management, training and education and other safety process management controls need to be continuously monitored and evaluated. This not only helps ensure the optimal effectiveness of a safety program but also serves to identify trends within the workplace (Manuele, 2017; OSHA, 2017a). OSHA maintains requirements to inspect workplace safety in the aviation community. No-notice and scheduled inspections are meant to confirm that standards are being followed, grievances are addressed, and overall workplace safety is being upheld (Jung & Makowsky, 2014).

Other Influencing Standards

SMS and ISO standards. The International Organization for Standardization (ISO) is a non-governmental entity dedicated to encouraging cross-national consensus on multiple items impacting the global population. ISO interfaces with governmental agencies across the globe to promote conformity and compatibility to normative processes and standards (ISO, 2017). The goal is to attempt to harmonize mandates, regulations, and laws for nearly all industries so they can operate openly and safely throughout the globe.

The ISO interfaces with various agency representatives to compare and contrast international legal concerns (ISO, 2017). This is particularly useful to the aviation community at large, as international carriers are less likely to be subject to another nation's legal confines in order to operate. The harmonization of safety protocol has propagated to other nation's aviation administrations (ISO, 2017). Likewise, other nations have influenced OSHA's rulemaking by sharing concerns and solutions through their ISO membership. Thus, while SMS, protocols, and plans may go by different nomenclature and codification, the context is nearly identical across the industry.

SMS and ANSI standards. The American National Standards Institute (ANSI) is a governmental accreditation body concerned with standardization and conformity. Consequently, ANSI is the United States member representative within ISO, giving voice to our nation's position concerning global industry safety matters. Their stated mission is to "enhance both the global competitiveness of the U.S. business and the U.S. quality of life by promoting and facilitating voluntary consensus standards and conformity assessment systems and safeguarding their integrity" (ANSI, 2017a).

ANSI could be considered the linkage between the international community and the United States concerning definitions, measures, proposed requirements, and mandated rules. As such, ANSI gives a voice to the national and international air industry to ensure the safety, security, and health of the aviation community (ANSI, 2017b). The openness and availability for interface presents a tremendous value to the safety management professional. Regular meetings, publications, and notes concerning international safety practices and concerns allow for an open dialog among all parties—government, industry, and international stakeholders (ANSI, 2017b). This state of affairs allows for proactive engagement and planning for the evolution of SMS and related standards.

SMS AND THE STANDARDS SUPPORTING IT

ANSI Z10

In July of 2005, the American Industrial Hygiene Association (AIHA) published the first version of the ANSI Z10 standard for occupational health and safety management systems (OHSMS). The ANSI Z10 standard was most recently updated in 2012. It was developed as a national standard through consensus-based collaboration among safety, health, and environmental professionals (Abrams, 2006).

The ANSI Z10 standard is generally compatible with the U.S. Occupational Health and Safety Administration (OSHA) regulations. Unlike OSHA regulations, the Z10 standard defines voluntary policy and process guidelines designed to promote the integration of safety into an organization's business management (Abrams, 2006). The Z10 standard combines elements of quality, environmental, safety, and risk management process guidelines to define management system standards for occupational health and safety. The ANSI Z10 standard is based on a compilation of widely accepted U.S. and international safety practices, along with well-established management systems concepts. The global adoption of these common safety practices sets a standard of care and creates a liability exposure for an organization, if its safety management practices and systems fail to address any related hazards.

Content description. The ANSI Z10 standard describes a systems-oriented approach to the management of safety and health in the workplace. It was formed during a six-year collaboration, discussion, and consensus-based effort by safety and health professionals from a variety of industry, government, academic, business, and labor organizations. The Z10 development committee and the numerous stakeholders who contributed as part of the effort were diligent about assessing and incorporating relevant safety material and management systems concepts from other standards, guidelines, and practices around the world. The resulting Z10 standard was designed to facilitate safety and health benefits while supporting the achievement of other business goals, such as improved productivity, quality, and profit returns (Manuele, 2008).

The widespread applicability and broad acceptance of the Z10 standard sets safety management norms organizations will be expected to meet. Though the scope of Z10 includes the standard as a set of minimum requirements for OHSMS, parts of the standard will be applied more comprehensively in some organizations than in others.

Management leadership and employee participation. The Z10 standard states that top management leadership and employee participation are critical for the success of an OHSMS (ANSI, 2012). Manuele (2008) found management involvement and active employee participation to be critically important in meeting safety culture goals and priorities for the organization. Top management must also define the expectations and provide visible commitment to the continuous maintenance and improvement of the OHSMS. The Z10 standard calls on top management to direct the organization to implement an effective OHSMS by "integrating health and safety into the daily functions of the business" (ANSI, 2012, p. E.3.1.1).

The successful implementation of SMS requires the following top management leadership actions:

a) Provide direction to the organization mandating the establishment, implementation, and maintenance of OHSMS per the Z10 standard.
b) Establish a documented policy to formalize top management's commitment to employee health and safety, employee participation, conformance to all safety policies and laws, and continuous improvement.

c) Accept the responsibility for implementing, maintaining, and monitoring the effectiveness of the organization's OHSMS.
d) Engage employees in formally defined roles to enable effective participation in the execution, maintenance, and operation of the OHSMS at all levels of the organization.

Top management bears the legal responsibility for safety and must ensure the organization adopts, executes, and abides by the regulations, laws, and accepted industry standards. Top management cannot implement, execute, and maintain the SMS and an effective safety culture without the engagement and participation of the employees at all levels of the organization (Friend, 2011). The safety execution vision and oversight function should be assigned to a top-level staff position reporting directly to the Chief Executive Officer (CEO) of the company. Though the safety department resides at nearly the top level of the organization, its expertise and knowledge pool are used to provide strategic guidance, purpose, and justification for the safety mission. The safety department should inspire and support the organizational management teams and employees so they can structure, implement, execute, maintain, and improve the organization's safety practices and SMS. "The key here is to involve as many employees as possible at the grass roots . . . those closest to the work are often in the best position to review it and to provide relevant feedback" (Friend, 2011, p. 105).

In summary, the chief safety officer must be part of the CEO's executive team and must have the authority to work with the CEO to ensure safety is an integral priority within the business strategy. The entire staff, at all levels of the organization, must have clearly defined, meaningful roles in the definition, implementation, assessment, and improvement of safety practices and the SMS.

Planning. Once the executive management team has made the commitment to prioritize safety within the organization and has directed the management team to engage the employees in the implementation of an OHSMS, the planning work may begin. Section 4.0 of the Z10 standard defines the requirements and describes the process for planning. The purpose of planning is to establish an organized approach for the identification, prioritization, management, and evaluation of OHSMS issues and their remedies. OHSMS issues include hazards, risks, management-system deficiencies, and opportunities for improvement (ANSI, 2012).

Z10 uses the Shewart cycle, also referred to as the Plan-Do-Check-Act (PDCA) cycle (Yantiss, 2011), as a model to provide an organized structure for the iterative sections of the safety management system standard (see

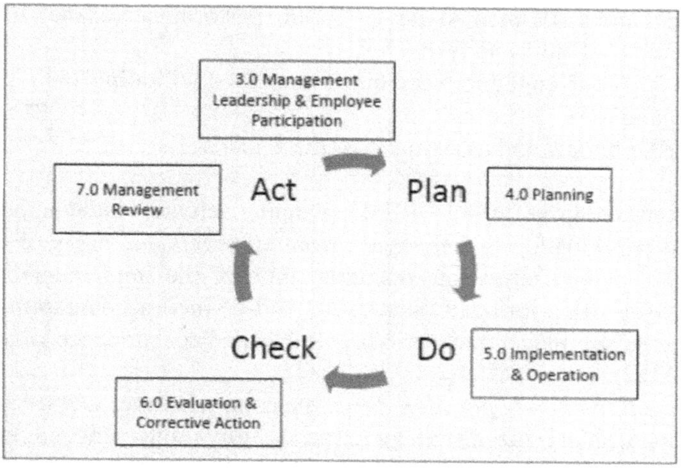

Figure 1.1 Shewart Cycle as Used by the ANSI Z10 Standard.

figure 1.1). Yantiss highlights the following situational objectives as strong candidates for the use of the PDCA cycle:

- continuous improvement and change management,
- product and/or process development,
- data collection for problem verification, prioritization, or root cause analysis, and
- implementation of change.

These objectives are all fundamental characteristics of safety management systems.

The planning effort in the SMS is a recurring process, as depicted in the PDCA cycle. When OHSMS is implemented, there are multiple types of ongoing planning as described in section 4.0 of the standard. Planning includes

1. the establishment of processes to identify OHSMS issues,
2. prioritization of OHSMS issues,
3. development of new objectives for the management system and for risk control,
4. implementation of strategies to address the prioritized issues and objectives,
5. follow up and management of prioritized issues and objectives,

6. review and assessment strategies for the performance of the OHSMS,
7. allocation of staff and resources,
8. strategies for engaging employees and for facilitating their participation, and
9. periodic review and assessment of these plans.

Key considerations during OHSMS planning activities include the engagement and participation of employees from all levels and areas of the organization, clear and up-to-date documentation of the implementation plans for achieving the prioritized objectives, well-defined accountability for the execution of the plans, and periodic review and maintenance of the plans (ANSI, 2012).

Implementation and operation. Once the comprehensive OHSMS plan has been defined, an organized strategy to successfully implement and operate the plan is needed. The Implementation and Operation section (5.0) of the Z10 standard provides a structure for using the OHSMS plan. This section constitutes the "do" phase of the PDCA cycle and it contains key requirements and considerations for the implementation and operation of the OHSMS plan. Though ANSI Z10 is a voluntary standard, designed to permit organizations to choose a business-relevant implementation strategy, it is important for users to comply. A poor implementation of safety management can result in potential liability hazards for the business.

A careful implementation of the Z10 standard will mitigate the potential for safety-related liability and systematically build the organization's safety management processes, capabilities, and culture in a continuous improvement-oriented manner (ANSI, 2012). To accomplish the mitigation of safety-related liability, while creating a system designed for continuous improvement, the following operational elements are required:

- Risk assessment processes.
- Prioritized methods of risk control targeting the elimination of hazards over actions designed to protect people from hazards.
- Processes to eliminate hazards during the design stage or change management effort, instead of accepting such hazards and attempting to protect humans from being injured by them.
- Procurement processes designed to minimize the introduction of potentially hazardous equipment, materials, goods, and services into the working environment.
- Management policies and processes to assess, control, and manage potential health and safety risks related to contractor work within the organization.
- Well-rehearsed response plans and processes to minimize safety and health risks related to emergency events.

- Education, training, awareness, and competence assessment processes to ensure all employees and contractors act in accordance with the OHSMS requirements, systems, and work practices.
- Communications processes to ensure employees and contractors are aware of the OHSMS implementation status and progress, as well as to ensure employees and contractors participate in the safety improvement and safety reporting programs.
- Documentation and record control processes to ensure the documents and records prescribed by the OHSMS are accurate, up to date, properly controlled, and adequately protected, so they will be accessible and usable when needed.

The ANSI Z10 standard describes these operational elements in greater detail and provides background information, tips, and examples to help with the design and implementation of the OHSMS. The list of operational elements may seem involved and lengthy at first; however, the task of implementing the processes from each of these areas is both essential and doable. It is important to consider all processes, as they work collectively to make the execution function, within the PDCA cycle, effective. Without a robust execution function, the OHSMS will likely fail to yield the expected business performance and cost-savings benefits.

If the magnitude of the implementation effort seems overwhelming, the organization may consider using an incremental change strategy. An incremental approach to change management can keep the workload manageable and provide frequent opportunities to celebrate progress (Yantiss, 2011). As the OHSMS deployment progresses and begins to impact the organization's approach to health and safety management, it is critical to evaluate and correct any under-performing elements in the system. Ignoring the best practices defined in the Z10 consensus standard can be analogous to violating the generally accepted, national standard of care regarding safety. This can lead to personal, professional, and corporate liability with costly lawsuits, increased insurance rates, and expensive fines (Abrams, 2006).

Evaluation and corrective action. Section 6.0 of the Z10 standard describes the requirements for evaluating the OHSMS plan and its operational performance. The OHSMS requires the monitoring and measurement of the performance of both the organization and the system. Organizational and system performance can be examined using audit processes, monitoring metrics, assessment techniques, and incident investigation analysis reports. The OHSMS procedures must ensure any identified corrective actions are properly executed, documented, and communicated.

In addition to removing hazards and achieving compliance with safety regulations, corrective actions must address systemic issues to prevent business processes and OHSMS-guided work practices from allowing hazards to develop (Manuele, 2006). Removal of systemic faults is accomplished using root-cause-analysis methods to examine the OHSMS performance data and to make improvements to the safety management plan. The findings and recommendations of this evaluation process must be communicated to the top management team so it can make high-level strategy improvements related to safety in the workplace. This completes the PDCA circuit and allows for continuous improvement in the organization's practice of safety management.

Management review. The successful implementation of safety practices, including the implementation of an OHSMS, is dependent upon the consistent commitment, support, and leadership from top management. Top management owns the legal responsibility for safety because it defines the organization's primary goals and priorities. Management must value, lead, and establish safety as a fundamental part of the workplace culture. Management must take responsibility for establishing the organizational structures, processes, tools, skills, and behavioral norms so that everyone is motivated and capable of competently supporting safety (Friend, 2011).

Section 7.0 of the Z10 standard calls for management to review the performance of the organization's OHSMS to lead the resolution of performance gaps and to act on opportunities for improvements. Management should review the performance of the OHSMS at least annually. The management review is the final key component of the PDCA cycle. It is an important part of the safety ownership role for the organization's top managers, as it provides them with the structure, engagement, and insight needed to fulfill their leadership responsibilities (Manuele, 2006).

ANSI Z10 deployment as a standard. ANSI Z10 is an established, national-consensus standard that continues to evolve and gain acceptance across a broad variety of industries. Its growing adoption and use by government, corporate, and educational organizations gives it increasing legal status as a standard of care for which top managers are likely to be held accountable. ANSI Z10 is not the only guidance available for safety management systems, however. It is important to note that other, more prescriptive, mandates for safety management systems also exist. One important example can be found in the aviation industry.

INTERNATIONAL SAFETY MANAGEMENT FOR CIVIL AVIATION

Until recently, safety management principles were provided primarily as voluntary guidance concepts only, much like the OHSAS 18001, AS/NZS 4801, and ANSI Z10. Without an international safety management standard, there will likely always be national variability within these concepts. Most are generally based on the concise PDCA concepts within ISO 9000 quality management criteria, as described earlier. The International Civil Aviation Organization (ICAO) has mandated consolidation of SMS principles into a nearly worldwide aviation standard to be incorporated into the legislation of their member nations. Safety management practitioners of all disciplines may be able to glean pertinent information from the methods ICAO used to design their international standard and implement it throughout the world.

Background and evolution. Aviation safety, not unlike other industrial safety programs, has evolved over the years by emphasizing different mishap-prevention strategies in accordance with the most prevalent threats. Just after the birth of aviation, most safety deficiencies were related to technical factors and failures of the aircraft themselves (ICAO, 2013a). As the frequency of technical mishaps decreased over the years, emphasis was then placed on regulatory compliance and human factors (ICAO, 2013a). Aviation safety has now evolved where "sound safety planning, including hazard identification, risk management and safety assurance, must be based on a thorough understanding of the processes and activities of people in the system, and the other components of the systems and environments in which they work" (Arendt & Adamski, 2011, p. 1). ICAO, established under the United Nations with jurisdiction over international civil aviation, has taken this new safety emphasis and, as of November 2013, made safety management a required standard for all 191 member nations. In addition to providing legislation, ICAO has also published several documents, training programs, and materials to assist its member nations and subordinate operators in developing their safety management programs. ICAO Order 9859, Safety Management Manual, was first introduced in 2006 and provides tools and guidance for safety management implementation in the aviation-related environment. It is readily available via online resources and is also an excellent reference for any safety professional seeking to implement a safety management process.

Content description. ICAO 9859 is composed of five safety management chapters along with several appendices and attachments. Chapter 1 provides an overview of the manual, while chapters 2 to 5 discuss safety management fundamentals; the relevant Standards and Recommended Practices (SARP); and then the phased development, implementation, and maintenance of safety management at the State Safety Program level (e.g., the United States Federal Aviation Administration, United Kingdom Civil Aviation Authority), as well as operations level where the safety management principles and guidelines transform into the SMS.

- The ICAO not only prescribed the standard of safety management within its legislation, but also explained *why* its implementation is important and *how* to implement utilizing fundamental concepts within Order 9859, chapter 2. Specific topics within chapter 2 include the following:

- The concept of safety.
- The evolution of safety.
- Accident causation.
- People, context, and safety.
- Error and violations.
- Safety culture.
- Management dilemma.
- Change management.
- Integration of management systems.
- Safety reporting and investigation.
- Safety data collection and analysis.
- Safety indicators and performance monitoring.
- Hazards.
- Safety risk.
- Safety risk management.
- Prescriptive and performance-based requirements.

The components outlined in chapter 2 of Order 9859 are universal to most organizations seeking to implement and apply safety management. Within aviation, this application begins with the State Safety Program, or in the case of the United States, the FAA's safety program legislation and administration. For industrial safety programs within the United States, this would be tantamount to ANSI Z10 being adopted as an OSHA mandatory standard. Each "State," or country, must establish the program infrastructure by providing overall specifications for performance, personnel, and processes within their organization, who is responsible for the implementation, and sustainment of the safety management program (ICAO, 2013). These compliance requirements include the enactment of legislation, regulations, policies, and directives to support the delivery of the product, in this case air transportation. As stated in ICAO Order 9859 (2013), chapter 4, the overall program should be scaled to the size and complexity of its civil aviation program and incorporate a mechanism for maintaining and evaluating its effectiveness based on the four safety management components: (1) safety policy and objectives, (2) safety risk management, (3) safety assurance, and (4) safety promotion. The compliance items within these four components are attributable to the national level program but also apply to the operators seeking to conform to the safety management requirements; therefore, the following descriptions apply to both entities.

Within the safety policy and objectives component, a state is expected to implement a safety legislative framework, outline the responsibilities and accountabilities regarding the establishment and maintenance of the safety management program, and establish an independent accident and incident investigation process and an enforcement policy outlining the conditions in which operators are allowed to internally manage safety deviations. Application and enforcement of these criteria may be different than other safety programs. In this case, states are expected to continuously monitor their safety management program at the operator level to ensure it has accounted for hazard identification, risk management, and the resulting risk controls. When the regulatory body inspects an operator as part of the validation process, it expects to see a self-sufficient safety program with documentation of deviations and emphasis on strengthening risk controls rather than punishing individuals. Most safety deviations are not a result of malevolence; therefore, these events should be used as an opportunity to learn and enhance the processes within the organization.

In a safety management program founded on the collection of verifiable data and records, the regulator validates that the operator's safety management program is abiding by national and international guidance and is properly managing risk through controls, identified in the second SMS component, safety risk management. These documents, in part, become a necessary component of the data-centric approach to safety.

Within the component of safety risk management, both state providers and operators should include items such as the establishment of controls governing how hazards will be identified and how safety risk will be managed. Operators should also identify program performance standards and conduct periodic reviews of the safety management system. This may appear relatively straight-forward; however, safety risk management is generally considered the backbone of any successful safety management program and must be given considerable attention at all levels of a safety program.

The third component, state safety assurance, primarily ensures safety oversight at the regulatory and operator level. This includes inspections, audits, and surveys to account for risk control integration, maintenance, and intended performance. Prioritization of these assurance functions should be given to the areas of greatest safety concern or items found to have the highest level of risk, as identified throughout the safety risk management process. Operators should expect the national-level regulatory body to establish a mechanism to collect and disseminate hazard identification and risk control best practices to other related organizations. It is worth noting this concept of collecting and disseminating information to agencies or organizations outside of one's own business may be foreign to operators who do not wish to share trade secrets. As it pertains to safety management, however, especially within the context of public air travel, the dissemination of safety information benefits the entire industry and its users. For instance, one particular airline may discover a procedural hazard with a new piece of installed avionics. This information, including their risk control process, may be extremely useful to other airlines utilizing the same technology. Operators within other disciplines can realize similar benefits by sharing safety management information through open communication channels. Effective communication is essential. not only within safety assurance, but also in the final safety management component, safety promotion.

Training, communication, and dissemination of safety information throughout an operator's organization falls under the domain of safety promotion. Considering the concepts within safety management and SMS may be unfamiliar, it is essential for the operator to provide the information necessary to facilitate an orderly transition to these new requirements. Good examples of safety promotion can be accessed via the ICAO safety management website at http://www.icao.int. There, a casual observer may find documents, training, and announcements pertinent to the safety management effort within international aviation and then apply the concepts to other disciplines. This promotion applies to the manner ICAO communicates with its state members, how organizations such as the FAA communicate with the various airlines and aircraft manufacturers, and how those businesses communicate with their entire staffs. Through effective communication and the other components of safety management, an operator's safety program should culminate in a pervasive safety culture perpetuating the implementation process. Operators should be aware it may take several years to effectively implement a new safety management program, dependent upon the size of the organization. ICAO recommends a phased approach to safety management implementation, as described within ICAO Order 9859. A phased approach is necessary to achieving a longstanding safety culture rather than simply building a program in name only. Thus, operators should be cautioned to not rush development.

The remaining chapter of ICAO Order 9859, namely chapter 5, Safety Management System, addresses ancillary systems to include other organizational activities such as finance, human resources, and legal departments. Due to this, it is important for an operator to consider all internal and external influences of their organization's system, preferably before large implementation efforts have been enacted. An operator intending to implement a safety management system may do well in selecting suppliers and subcontractors who already utilize this type of safety approach.

Implementation. ICAO legislation consolidated aviation safety management principles and promoted them to a mandatory standard for ICAO's 191 Member States accounting for nearly all of the world's independent countries. Operators within other disciplines may be able to learn from ICAO's publicly available safety management implementation documents. Operators may also learn from organizations such as the FAA, who sought to combine existing programs with new ICAO requirements. The FAA previously abided by ISO 9000 quality management standards and already oversaw a robust risk management program. The new safety management standards of ICAO did not require the FAA to create several new programs but only required they bolster areas of deficiency and maintain areas of superiority. Operators implementing safety management are likely to face similar situations. The interested reader is directed toward documents such as FAA Order 8000.369A (2013a), Safety Management System; Order VS 8000.370A (2013b), Aviation Safety (AVS) Safety Policy, and Order 8000.367A (2012), AVS Safety Management System Requirement, in addition to the FAA's SMS implementation and resource website for more information.

Standard of care considerations. The international aviation safety management program has its roots in ICAO's voluntary recommended practices and is now considered mandatory for its Member States. Over the past several years, the U.S., Canada, and other countries have developed voluntary, aviation SMS in anticipation of mandatory guidance. While changes to their established programs will be necessary to fully comply with ICAO mandates, the transparent communication between ICAO and its associated nations has allowed the majority of the SMS infrastructure to be built within countries that have chosen to be proactive. Countries waiting for final guidance before SMS implementation will have these templates and expertise available to construct their own aviation SMS as the majority of the world transitions to this international aviation safety management standard. Operators of a variety of other disciplines are encouraged to consider ICAO's transition from voluntary to mandatory standards. While ISO cannot mandate an operator to adhere to an international standard, national organizations such as the U.S. Occupational Health and Safety Administration, the European Agency for Safety and Health at Work, and the Canadian Centre for Occupational Health and Safety may eventually transition their voluntary standards to mandatory ones.

Considerations for Selecting Appropriate Safety Management Standards

The ICAO and ANSI Z10 standards define similar requirements for implementing and operating a safety management system. Other safety management system standards exist for specific countries and specific industries. The British Standard Occupation Health and Safety Assessment Specification (OHSAS) 18001 and the Australian/New Zealand Standard 4801 are examples of country-specific SMS standards. The aviation, maritime, and railway industries all have one or more sets of requirements for the implementation and operation of an industry-relevant SMS.

Applicability to Organizational Mission and Operational Requirements

In order to achieve sustained success, the implementation and operation of the SMS must support the organization's mission, goals, and operational requirements. If the requirements and processes of the SMS conflict with the organization's mission, goals, and values, the actions and culture required for consistent safety in the workplace will not evolve and prevail. It is essential to implement SMS strategies appropriate for the organization's size, work activities, and risk exposures. Aviation-oriented companies will likely be best served by aviation-oriented standards. Industrial organizations should likely seek those standards that target them. If additional SMS processes or safety features from other standards will improve safety management effectiveness, those features or processes should also be implemented into the SMS.

When conflicts are encountered between the organization's mission and SMS requirements, top management action will be needed to assess the relevance of the mission goals against the systemic importance of the SMS requirements. Some SMS requirements may have insignificant impacts on the management of safety within a given organizational context. In other cases, the organization's mission may need to be modified to incorporate the productivity and financial benefits associated with employee health and safety-oriented business goals. Regardless of where the adaptation takes place, the organizational mission, business goals, and SMS implementation requirements must become aligned. In most cases, the simplest approach to achieving this alignment will be to start with the standard best suited to the organization's size, mission, workplace environment, and health and safety risk challenges.

FAA GUIDANCE

The focus the aviation industry has placed on safety has paid immense dividends. Safety Management Systems are the current state-of-the-art in improving safety. SMS has proven so successful that the U.S. air carriers are required to use them (14 CFR Part 5). The ICAO requires participating nations to establish a State Safety Programme (SSP) to oversee the SMSwithin the State's jurisdiction. The following sections will examine the FAA's SSP, the good, bad, and ugly, with respect to the linkages between the FAA's SSP and the FAA-mandated SMS. The FAA's most recent guidance to organizations on SMS is found in Advisory Circular (AC) 120-92B, *Introduction to Safety Management Systems for Aviation Service Providers*. The actual regulatory requirements are found in Title 14 of the Code of Federal Regulations, Part 5, *Safety Management Systems*. The FAA's SSP is found in FAA Order 8000.369A, *Safety Management System.*

How the FAA Expects Air Service Provider (ASP) SMS to Work: AC 120-92B

AC 120-92B, released in 2015, gives Part 121 air carriers, who are required have an established SMS, a description of the regulatory requirements, guidance, and methods of developing and implementing SMS (FAA, 2015). The FAA also recommends AC 120-92B for any service provider seeking to voluntarily establish their own SMS. The FAA has, in AC 120-92B, provided information explaining the concepts and processes the SMS should embody. AC 120-92B clearly outlines the regulatory requirements found in Part 5, provides explanations on how each of the processes integrates into the SMS, and even discusses how to apply the process to small, medium, and large organizations (FAA, 2015). A very thorough chapter is provided explaining how to implement SMS from mapping the organization, conducting a gap analysis, and developing an implementation plan with a recommended four-phase approach.

Regulatory Requirements: 14 CFR Part 5

Part 5 is a short section of the aviation regulations that establishes the requirements of SMS for operators required to have one. This regulation directs organizations to establish and document safety policies and conduct a thorough risk management process. Organizations must create and maintain systems to monitor the safety process through collecting and analyzing appropriate data. Organizations are required to conduct assessments to verify adherence to and efficacy of established risk controls, as well as to establish and maintain a continuous improvement process for the SMS. Additionally, the regulations require employee training on the safety system.

How the FAA SMS (SSP) Should Work: FAA Order 8000.369A

According to FAA Order 8000.369A, SMS is expected to meet most of the requirements of ICAO's SSP framework (FAA, 2013). Similar to the AC, the Order provides the statutory authority governing the SSP and describes the four components of policy, risk management, assurance, and promotion. The FAA's role, when dealing with organizations within the aerospace system, is to "ensure the SMS is comprehensive and functioning properly to enable the aviation product/service provider to appropriately manage its safety risk" (FAA, 2013, p. 10).

The FAA uses rules and regulations, policy, and standards as its primary way of providing oversight (FAA, 2013). With limited personnel, the FAA cannot monitor every operation at all times. Instead, it establishes the limits to which most operators conform. Violators are penalized through fines and certificate actions. The FAA utilizes oversight and surveillance activities to fulfill oversight responsibilities.

> "FAA oversight personnel conduct inspections, audits, and evaluations to provide objective evidence to assess product/service provider process and product design adequacy, the continued operational safety of its product or service, and the performance of its SMS" (FAA, 2013, p. 10).

FAA Oversight: Hooks and Links

AC 120-92B, 14 CFR Part 5, and Order 8000.369A all work together to define the overall national aviation safety scheme. It is within the interrelationships of these documents that the FAA establishes oversight of the SMS within its purview. These documents provide guidance that varies from specific to vague. The Advisory Circular by nature provides guidance, but not direction, in complying with the regulations. The 14 CFR Part 5 details what is required in the SMS, but does not prescribe how to go about meeting those requirements. Even FAA Order 800-369A provides broad guidance that is applicable to the FAA, as a whole, rather than providing specific details. This lack of specificity in both FAA Order 8000.369A and AC 120-92B is a double-edged sword. The intent is to allow the flexibility to craft programs tailored to the specific need of each unique organization. The FAA recognizes there is no single method of compliance that will work in every situation or for every operator (FAA, 2015). While this flexibility is laudable, it also creates room for misinterpretation and misunderstanding in application. The confusion could lead to potential breakdowns in the mechanisms, the hooks and links, which are essential to proper and adequate oversight by the FAA.

How the FAA SMS and ASP SMS Interconnect?

Hooks. For the purposes of this text, the term "hooks" refers to the regulatory requirements contained in 14 CFR Part 5. The regulation defines the minimum acceptable performance to fulfill the requirements. Failure to achieve and maintain this level of performance can result in adverse action against an operator. This can include monetary fines or the suspension or revocation of the operator's certificate.

Links. The term "links" refers to the interfaces between the FAA and the organizations operating under the FAA's purview. They are the activities and functions that the FAA uses to ensure compliance with the regulatory hooks contained in 14 CFR Part 5. Cumulatively, the links are the outputs of the various safety processes within the operators' SMSs. The guidance within AC 120-92B and Order 8000.369A clearly demonstrates the connection between the FAA's responsibility for oversight and the required performance by the operators. If the guidance provided by the FAA is thoroughly followed, the links will be in alignment.

Alignment. It would be expected that many of these hooks and links would be well-aligned to facilitate the desired oversight function. In other cases, there may be some misalignment impairing the FAA's ability to perform adequate oversight. There is enough ambiguity between these two guidance documents allowing for misdirection and misunderstanding. Thus, the links will be misaligned. An examination of these hooks and links will help to determine where trouble may lie.

Well aligned. A well-aligned linkage exists between the FAA and the ASP when the hooks set by 14 CFR Part 5 and the links within the recommendations for ASPs in AC 120-92B and the FAA's internal guidance in Order 8000.369A are matched in a way that they can support each other. For example, in order for the FAA to provide SMS oversight, the SMS must be in place. 14 CFR Part 5 § 5.1 specifies that air carriers operating under 14 CFR Part 121 (scheduled airlines) must have an established SMS acceptable to the FAA in place not later than March 9, 2018. To meet this timeline, these operators must have submitted an implementation plan by September 9, 2015, which would have needed to be approved by March 9, 2016 (FAA, 2001). This is a very clear timeline operators must follow to comply with regulations sets a hook. AC 120-92B (2015) provides the operators with expanded information to assist in complying with the requirements of 14 CFR Part 5. As with all Advisory Circulars, it provides an acceptable means of compliance, but strict adherence to the AC is not mandated as long as the requirements of the regulation are met. FAA Order 8000.369A outlines the FAA's responsibility in evaluating an operator's SMS as part of its oversight responsibilities (2013). This SMS mandate establishes the first set of hooks and links.

The components of SMS must include are outlined in 14 CFR Part 5 § 5.3 (2010). Subparts B through F of Part 5 outline the requirements for safety policy, safety risk management, safety assurance, safety promotion, and record keeping. The next important set of hooks and links is found in safety assurance and record keeping. Safety Assurance requirements within Part 5 mandate the operator have processes in place to collect safety performance data. This data is critical to the effectiveness of FAA audits. In addition, the data collected across the industry will provide the FAA with the ability to perform trend analysis on a national scale. Under 14 CFR Part 5 § 5.97 operators are required to maintain records of the outputs of the various SMS components for specified time periods. This gives the FAA the opportunity to review the data periodically to evaluate the health of the operator's safety system. AC 120-92B explains in further detail what types of data must be collected and stored, as well as recommendations regarding how to perform data analysis. FAA Order 8000.369A discusses the FAA's requirement to perform audits, inspections, and evaluations of operators. Reviewing the collected safety performance data is a critical component within the FAA's auditing process. Data collection and storage is another hook-and-link set.

Misaligned links. A misaligned linkage exists between the FAA and the ASP when the hooks set by 14 CFR Part 5 and the links within the recommendations for ASPs in AC 120-92B (2015) and the FAA's internal guidance in Order 8000.369A (2013) lack a clear relationship. The overall structure of the two documents exhibits this lack of alignment. AC 120-92B, while not directive, provides the service provider with great clarity as to what is required to meet the intent of 14 CFR Part 5. In addition, the AC also provides a wealth of amplifying information regarding each component of an FAA-compliant SMS. The AC takes each section of 14 CFR Part 5 and explains the specific requirements of the regulation, as well as listing any related regulatory requirements. It also provides a plain language summary of the component and an explanation of the objective, along with the pertinent section of 14 CFR Part 5 shown verbatim. If necessary, a more detailed explanation of the requirement is provided, including recommendations on designing a compliant process. The AC even includes details on how the requirement could be scaled for small, medium, and large service providers.

By contrast, Order 8000.369A (2013) is very limited on details. This document provides brief descriptions of the components of a compliant SMS but without the level of detail or support contained in the AC. Overall, the Order is written with a very broad brush. There is so little specificity within the Order that there is an opportunity for misunderstanding and misinterpretation. This creates a significant misalignment in what should be a strong link.

CONCLUSION

There are many regulations to be considered by an aviation operator, large or small, based on OSHA oversight. The goal of the agency is to ensure and enforce a safe and healthy workplace on behalf of all employees. Voluntary safety awareness and training have been a hallmark of OSHA in the propagation of industry-wide safe cultures. By allowing specific tailoring conjoined with agency guidance, SMS tools and techniques have ensured the safety of both the public and industry professionals.

When designing and managing a robust safety program, a variety of factors must be considered. Planning, organization, and leadership skills are critical for the initiation and sustainment of any program. These skills become evident in any strong safety management system ensuring compliance with safety regulations and guidelines. Although mishaps may not be totally eliminated, theeffective SMS seeks to identify hazards and risks before they occur. When the unforeseen happens, it is incumbent upon the organization to have a response plan in place to ensure similar circumstances do not recur. Management of these processes can be expensive; therefore, budget considerations must be realistic and line up with the goals of minimizing risks and responding to incidents.

The power of a well-designed SMS to provide safer operations and work environments is surpassed only by keeping the planes locked in the hangar. In aviation, providing safe operations is crucial. ICAO has recognized the benefits to safety SMS provides and has recommended all member states develop state safety plans that mandate the development and use of SMS for commercial operations. To ensure that the U.S. air carriers of all sizes are adopting quality, well-designed SMS, the FAA has provided extensive guidance in designing and implementing an effective SMS. FAA Order 800.369A, AC 120-92B, and 14 CFR Part 5 are the three documents critical to SMS development, implementation, and continuous monitoring within the United States

REFERENCES

Abrams, A. L. (2006). Legal perspectives on ANSI Z10-2005. *The Compass, Special Edition, a Publication of the American Society of Safety Engineers.* Retrieved from www.asse.org/practicespecialties/docs/Z10 SpecialIssue.pdf.

American National Standards Institute (ANSI). (2012). *American National Standard – Occupational Health and Safety Management Systems* (ANSI/AIHA Z10-2012). Falls Church, VA: American Industrial Hygiene Association.

ANSI. (2017a). *About ANSI.* Retrieved from https://www.ansi.org/about_ansi/overview/overview?menuid=1.

ANSI. (2017b). *Introduction to ANSI.* Retrieved from https://www.ansi.org/about_ansi/introduction/introduction?menuid=1.

Arendt, D., & Adamski, A. (2011). System and task analysis. In A. J. Stolzer, J. J. Goglia, & C. D. Halford (Eds.), *Implementation of Safety Management Systems in Aviation* (pp. 1–51). Farnham, Surrey, GBR: Ashgate Publishing Group.

FAA. (2010). Safety management systems for aviation service providers. *Advisory Circular 120-92a.* Retrieved from https://www.faa.gov/documentLibrary/media/Advisory_Circular/AC%20120-92A.pdf.

FAA. (2012). Aviation safety (AVS) safety management system requirements. *FAA Order VS. 8000.367A.* Retrieved from https://www.faa.gov/documentLibrary/media/Order/FAA_Order_8000.367A.pdf.

FAA. (2013). Safety management system. *Order 8000.369A.* Retrieved from http://www.faa.gov/documentLibrary/media/Order/8000.369A.pdf.

FAA. (2015). Introduction to safety management systems for aviation service providers. *Advisory Circular 120-92B.* Retrieved from http://www.faa.gov/documentLibrary/media/Advisory_Circular/AC_120-92B.pdf.

FAA. (2017). Overview. *Title 14 of the Code of Federal Regulations (CFR).* Retrieved from https://www.faa.gov/regulations_policies/handbooks_manuals/aircraft/amt_handbook/media/FAA-8083-30_Ch12.pdf.

Federal Aviation Administration (FAA). (2010). 14 CFR Part 5- Safety management systems. *Electronic Code of Federal Regulations.* Retrieved from http://www.ecfr.gov/cgi-bin/retrieveECFR?gp=&SID=bf805515603efe479e21d23cdf7b9086&mc=true&n=pt14.1.5&r= PART&ty=HTML #sp14.1.5.a.

Friend, M. A. (2011). Top management support. In A. J. Stolzer, J. J. Goglia, & C. D. Halford (Eds.), *Implementation of Safety Management Systems in Aviation* (pp. 1–51). Farnham, Surrey, GBR: Ashgate Publishing Group.

International Civil Aviation Organization (ICAO). (2013a). *Safety Management Manual* (3rd ed.). Retrieved from http://www.icao.int/safety/SafetyManagement/Documents/Doc.9859.3rd %20 Edition.alltext.en.pdf.

ICAO. (2013b). *Safety Management Manual* (SMM) (3rd ed.). ICAO Order No. 9859. Retrieved from https://www.icao.int/safety/SafetyManagement/Documents/Doc.9859.3rd%20Edition.alltext.en.pdf.

ICAO. (2013c). Annex 19 safety management (1st ed.). *Integrated Safety Management.* Retrieved from http://cfapp.icao.int/tools/SMikit/story_content/external_files/Annex%2019%20-%20DETAILED%20PRESENTATION.pdf.

ICAO. (2013d). SMM 3rd edition highlights (Doc 9859). *Integrated Safety Management.* Retrieved from http://cfapp.icao.int/tools/SMikit/story_content/external_files/SMM%203rd%20EDITION%20HIGHLIGHTS.pdf.

International Organization for Standardization [ISO]. (2017). What are standards? *All About ISO.* Retrieved from https://www.iso.org/about-us.html.

Jung, J., & Makowsky, M. D. (2014). The determinants of federal and state enforcement of workplace safety regulations: OSHA inspections 1990–2010. *Journal of Regulatory Economics, 45*(1), 1–33.

Manuele, F. A. (2006, February). ANSI/AIHA Z10-2005: The new benchmark for safety management systems. *Professional Safety*, *25*, 25–33.

Manuele, F. A. (2008). *Advanced Safety Management Focusing on Z10 and Serious Injury Prevention.* Hoboken, NJ: John Wiley & Sons.

Manuele, F. A. (2017). OSHA recommended practices for safety & health programs. *Professional Safety, 62*(3), 32.

Occupational Safety and Health Administration [OSHA]. (2015). *OSHA Safety and Health Program Management Guidelines.* Retrieved from https://www.osha.gov/shpmguidelines/SHPM_guidelines.pdf.

OSHA. (2017a). About OSHA. *United States Department of Labor.* Retrieved from https://www.osha.gov/about.html.

OSHA. (2017b). All about VPP. *United States Department of Labor.* Retrieved from https://www.osha.gov/dcsp/vpp/all_about_vpp.html.

OSHA. (2019). OSHA penalties. *United States Department of Labor.* Retrieved from https://www.osha.gov/penalties/.

Schmidt, P. (2002). Pursuing regulatory relief: Strategic participation and litigation in U.S. OSHA rulemaking. *Business and Politics, 4*(1), 71–89.

Yantiss, B. (2011). SMS implementation. In A. J. Stolzer, J. J. Goglia, & C. D. Halford (Eds.), *Implementation of Safety Management Systems in Aviation* (pp. 161–267). Burlington, VT: Ashgate.

Chapter 2

Designing and Managing the Safety System

Bradley S. Baugh

> What makes a good river guide great? An extensive tool kit. I'm not talking about a box with wrenches and screwdrivers. Great guides have solutions for almost all issues, be it gear, guest, or guide related. Kate Wollney, the Queen of the Rogue River in Oregon, is one of the best problem solvers around. She never gets frazzled or overwhelmed in the slightest. She attends to each snafu like it was part of the original plan, and keeps moving forward with the trip. (OARS, n.d.)

The FAA has mandated SMS to certain users and highly encouraged it for all other aviation users. As pointed out in FAA Advisory Circular 120-92B, the "what" of a program is specified, but the "how" is not, SMS is built upon the four components of safety policy, risk management, assurance, and safety promotion (FAA, 2015). The question thus remains, what tools are used to build upon these components? There is not a "one size fits all" application in building and maintaining the SMS. The result is what matters. With this in mind, the tools presented in this chapter come from the aviation industry and industries. This chapter provides an overview of common system safety tools and their practical application. It is divided into four parts: reactive tools, proactive tools, predictive tools, and best practices/lessons learned (from nonaviation-specific sources).

The four parts introduced above provide a framework for categorizing suggestions, good ideas, and best practices and are not intended to limit a particular type of tool deemed valuable. There seems to be an aviation-industry focus on a few popularly used tools outlined in government and professional literature. However, the savvy safety practitioner is always on the lookout for those tools that fit the organization and the task at hand. Descriptions of these "aviation" tools can be found in FAA, ICAO, and military sources. Other organizations, such as the ISO and the Project Management Institute (PMI), are also sources of tools directly applicable to aviation SMS. These organizations publish regulatory orders and industry standards in their particular sphere and offer basic tools for developing and managing the SMS. A search of peer-reviewed scholarly journals also provides practical advice. Some examples include works published in *Reliability Engineering & System Safety* (Favarò & Saleh, 2016), *Safety Science* (Donaldson, Borys, & Finch, 2013), and *Science International* (Hakeem-Ur-Rehman & Aslam, 2013).

"Tools," "methods," and "programs" are terms sometimes used synonymously. While this may lead to confusion, the important point is to grasp the concepts and when to use them for the benefit of the SMS. Another key point is that the aviation safety practitioner may be required to take on broader safety responsibilities for the organization in the industrial or occupational safety areas, so it is helpful to know a varied array of applicable tools. This chapter will provide practical examples of some of the tools using non-aviation examples to show greater applicability in the safety realm. ICAO defined "three methodologies for identifying hazards" (ICAO, 2013, pp. 2–260): reactive, proactive, and predictive. Tools are the instruments, or the "how-to" part of identifying the hazards under the three categories. The tools included below are not an exhaustive listing, and they range in complexity. Many tools have utility in multiple areas, so it is a best practice to periodically look "outside the box."

TOOLS FOR THE SMS TOOLBOX

During the opening stage of the 2017 Tirreno-Adriatico, a Union Cycliste Internationale World Tour professional cycling race, Team Sky suffered a major setback when, during the team time trial, several PRO wheels failed, in one case cause causing cyclist Gianni Moscon to crash (PRO is Shimano's performance component subsidiary). In a statement from Shimano, "PRO is continuing its investigation. . . . We are continuing to look closely into all factors that could cause the incident. . . . During production the three-spoke wheel passed PRO's extremely high internal quality control and ISO/UCI standards" (Evans, 2017). In June 2017, PRO stated,

> [We] conducted a thorough investigation to replicate the circumstances, to examine its strict safety standards and to ensure that the safety of riders remains the highest priority. The investigation concluded that the 3-spoke wheel is completely safe to continue racing. The incident occurred because the rider hit a pot hole at high speed on a tyre that did not meet the minimum tyre width specification as advised by PRO. (2017)

Reactive Tools

Traditionally, safety programs have in large part been based on analysis of the last mishap as a means of preventing the next mishap. Using the ICAO definition, reactive system safety involves "past outcomes or events" (2013, pp. 2–26). Reactive methodology seeks to find causes, so preventative measures can be implemented. Two major groups of tools are typically used in reactive safety management: data gathering and data analysis.

Data gathering tools. In a broad sense, these tools are how the information required to analyze incidents and mishaps and to provide a foundation for instituting corrective actions and preventative measures is gathered.

Mishap investigations. This is perhaps the broadest and most commonly thought of tool in the safety program toolbox. In another sense, the mishap investigation is a tool comprised of many other tools. There are many useful sources to help develop and tailor this tool to the organization; for example, ICAO (n.d.), National Transportation Safety Board (n.d.), National Safety Council (2017), Occupational Safety and Health Administration (n.d.), and the U.S. Department of Defense (U.S. Air Force, 2016; U.S. Army, 2015; U.S. Navy, 2005). The methods of conducting investigations vary, but the key outcome of the investigation is the data gathered during the investigation.

Interviews. The interview is a major tool in the mishap investigation toolbox. Sources on how to conduct a mishap interview are scarce. However, there is a clear need for solid communication and interpersonal skills. Two sources to help guide the development of this tool are the FAA's *Compliance and Enforcement* publication (2015) and *Investigative Interviewing Techniques: The PEACE Model*, from i-Site.com (2017). Whether information is gathered from a witness or a subject matter expert, the interview can help the investigator capture vital evidence.

Mishap plans. The midst of a crisis is not the time to figure out how to tackle the challenge of gathering the mishap data. A well-thought-out plan provides the outline for the investigation, speeds up the process, and allows for a better end product. If not immediately captured, some data is rapidly lost such as witness statements, ATC recordings, and ground scars. Time is always of the essence. Plans assist in identifying who does what and when, in order to ensure the data is available for the investigation. One resource for developing mishap plans is the U.S. Department of the Interior's Office of Aviation Services. Their *Interagency Aviation Mishap Response Guide & Checklist* (2014) can be used as a template for writing a plan.

Checklists. Checklists are a staple of emergency response actions in aviation. The best investigators use checklists to keep themselves safe in the mishap area and enable methodical data gathering.

Data analysis tools. The analysis tools are used to transform the data gathered from an incident or mishap into fix-action and mishap prevention plans. This is how investigators search for, and hopefully identify, causes. Using the data analysis tools, they are able to create the accident report.

Human Factors Analysis and Classification System (HFACS). In 1990, James Reason published his work, detailing what is now known as the "Swiss cheese" accident causation model. He explained, while organizations have many levels of safety defenses, latent errors cause holes in the defenses. When the holes line up, a mishap occurs (Reason, 1990). Building on this concept, Wiegmann and Shappell developed a model to define the latent errors in a framework usable to practitioners. The result of their work is the HFACS model. HFACS uses "four levels of failure: 1) Unsafe Acts, 2) Preconditions for Unsafe Acts, 3) Unsafe Supervision, and 4) Organizational Influence" (Wiegmann & Shappell, 2003, p. 50). This framework guides investigators to uncover underlying causes beyond "pilot error."

Root cause analysis. This tool can be useful in the mishap investigation and accident report analysis. Beginning with the mishap, investigators progressively analyze preceding actions by breaking them into smaller components until they arrive at the "but for this, the mishap would not have occurred" condition or hazard. Root cause analysis has long been associated with quality improvement processes. The methodology is the same, and therefore, organizations discussing quality make valuable sources. For example, the American Society for Quality has developed a primer on root cause analysis, its uses, and techniques to follow when conducting an analysis (2017).

Cause-and-effect diagrams. This tool is also known as the Ishikawa (after the developer), or fishbone diagram. The diagram graphically links mishap causes with known effects. This is another tool common to quality professionals, adaptable for the safety professional. It is still a type of root cause analysis, although, it is conducted with a different methodology. A sample cause-and-effect diagram is displayed at figure 2.1.

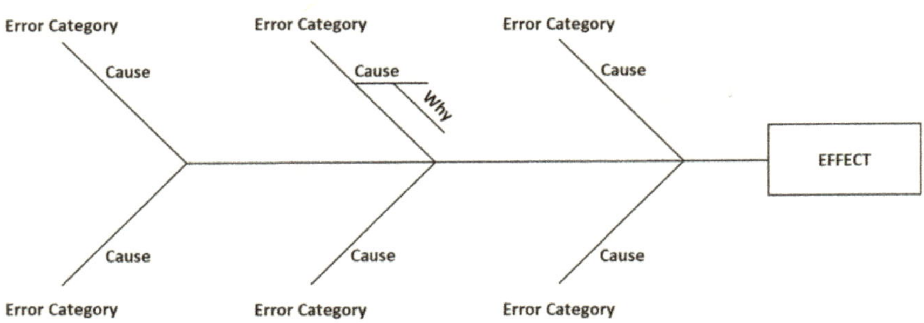

Figure 2.1 The Cause-and-Effect (Fishbone Diagram).

Accident report analysis. Part of reactive safety management includes analyzing reports from sources outside of the organization. When included as a component of an SMS, accident report analysis is systematically integrated with applicable lessons identified. Lessons learned can then be incorporated into mishap-prevention activities. This is an important tool, because it is better to learn from someone else's mistake when possible.

Proactive Tools

A better approach to preventing the next mishap is to develop programs and processes to proactively promote safety, by identifying precursor activities and stopping them prior to a mishap. If the near-miss is captured, the knowledge gained can be used to develop protective measures. In the ICAO definition, *proactive system safety* concentrates on the "analysis of existing or real-time situation[s]" (2013, pp. 2–26). Proactive tools assist in actively searching for and reporting discrepancies to prevent mishaps. The tools below are divided into three groups: data gathering tools, data analysis tools, and safety assurance tools.

Data gathering tools. There are many tool, methods, and programs developed to gather discrepancy data. Some are mandatory, based on the FAA operating certificate. Others are voluntary and provide incentives for self-reporting discrepancies, even if the discrepancy is a violation of federal regulations.

Aviation Safety Action Program (ASAP). This program tool encourages employees to voluntarily report safety problems, even if they are in violation of flight regulations. The tool was developed between certificate holders (Part 135 airlines, for instance) and the FAA through a Memorandum of Understanding (MOU). Federal Law protects ASAP reports from being disclosed to the public and protects the reporter from disciplinary action, as long as there was no willful intent (FAA, 2002). The goal of this tool is to capture data not always detected, without fear of adverse repercussions.

Flight Operational Quality Assurance (FOQA). A voluntary program at the federal level, FOQA is a data-aggregation mechanism collecting de-identified safety information, allowing the FAA to track national trends (FAA, 2004). Many organizations also have a FOQA at the company level, to mirror the FAA program and more closely focus on their operations.

Interviews. When used as a proactive tool, an interview helps safety and line managers encourage safety reporting. Often, it is the line operator who detects problems, yet sometimes findings are not reported. Periodic interviews can help break down the reporting barriers, so solutions can be developed.

Expert judgment. Expertise is found at all levels, from the line operators to upper levels of the organization. The safety manager exploits this tool

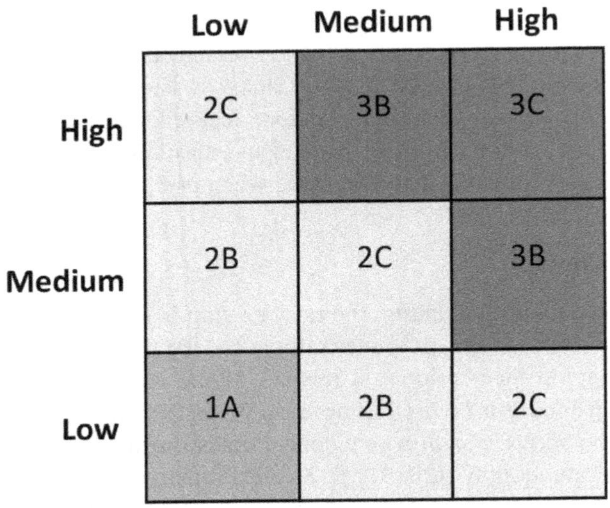

Figure 2.2 Illustrative Risk Matrix.

proactively by seeking out experts to augment current safety knowledge. This is often accomplished through interviewing those closest to the operation.

Data analysis tools. Proactive, data analysis tools begin where the reactive data analysis tools end. The tools analyze (or theorize) a condition prior to the event, in order to develop mitigation strategies.

Risk matrix. The risk matrix, also known as a probability and impact matrix, is a common tool in many industries and is tailored in complexity to the needs of the organization. A simple 3 × 3 risk matrix is displayed in figure 2.2.

Risk matrices are developed with the likelihood of a mishap (probabilities) on one axis and the severity of a mishap (impacts) on the other axis. The boxes can be assigned a score and/or colors, according to the organization's risk tolerances and policies. This score forms a basis for a safety risk index and helps visualize safety risk tolerability (ICAO, 2013). The efficacy of the risk matrix is debated; however, the risk matrix can help "paint a picture" for senior leaders, and the graphical nature should not be discounted.

Threat and Error Management (TEM). TEM is a framework to focus analysis on threats, errors, and undesired states. Its purpose is to "assist in understanding, from an operational perspective, the inter-relationship between safety and human performance in dynamic and challenging operational contexts" (ICAO, 2005, p. 2). Threats in this model are those things outside the control of the line operator, such as an increase in workload or complexity. Errors are those commissions or omissions leading to deviations. Undesired states are unsafe conditions (ICAO, 2005). Identifying the unsafe conditions allows for the "management" part of the TEM model.

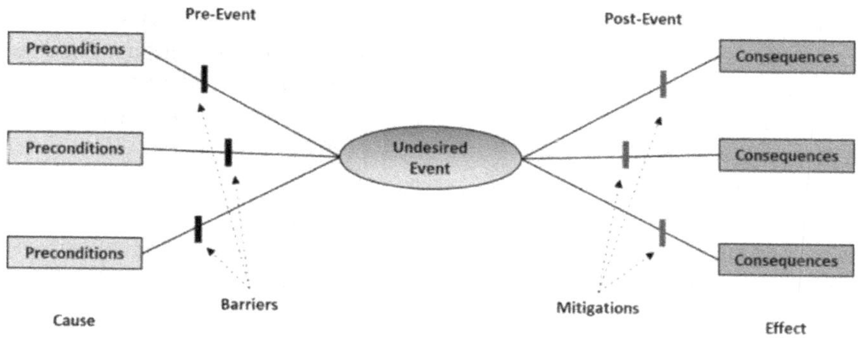

Figure 2.3 Bow-Tie Diagram.

Bow-tie diagram. This type of diagramming is useful for graphically depicting causes and effects. It also aids in visualizing barriers to stop the initiating events (listed as preconditions in the diagram) and mitigations to reduce the logical consequences. The undesired event is displayed in the middle of the diagram with the initiating events feeding into the event (if not blocked), and the consequences flowing out (if not mitigated). This can be seen in figure 2.3.

Brainstorming. This tool is used in many instances but is included here for its value in identifying risks and potential solutions. It is used in coordination with expert judgment, though "experts" are not necessarily required. Sometimes, guided brainstorming with solid line operators will provide inputs and solutions not found using other methods.

Safety assurance tools. The tools introduced here are included because of their interplay with proactive safety management; however, they also play a role in the overall SMS.

Inspections. Inspections take many forms and are performed at many levels. Some organizations use a self-inspection framework, while others have inspection organizations. They can take the form of a comprehensive assessment or a limited spot-check. Sometimes the tools and methods overlap. The key is reviewing actual performance for expected (regulated) performance, identifying gaps, and providing data in order to fix problems.

Internal Evaluation Program (IEP). The IEP is a voluntary tool the FAA especially encourages 14 CFR Part 121 and Part 135 operators to embrace. It is essentially a system process for self-evaluation and continuous improvement, specifically focused on regulatory compliance (FAA, 2006a).

Line Operations Safety Assessment (LOSA). This is a program of peer-to-peer observations. The observations are fed into the SMS to identify negative trends, system weaknesses, and system strengths (FAA, 2006b).

Voluntary Disclosure Reporting Program (VDRP). This FAA-level program provides a framework for self-reporting regulatory violations. The FAA will typically waive civil penalties for organizations that identify, promptly report, and aggressively fix violations (FAA, 2009). This tool enables other tools, such as the ASAP and IEP, to operate effectively,

Check sheet. Perhaps one of the most basic tools, the check sheet, is a list an inspector uses to record observations. It can be a simple count of certain observations or a complex matrix. Although the check sheet seems simple, it provides a method of systematically recording data needed to make safety decisions.

Predictive Tools

A more advanced way of preventing mishaps involves looking forward to prevent occurrences before there are even precursor events. It can be thought of as "horizon scanning" or trying to discover what might be the next mishap. Predictive system safety "involves data gathering. . . , analyzing system processes. . . , and the environment to identify potential future hazards and initiating mitigating actions" (ICAO, 2013, pp. 2–26). The tools here are not sub-categorized as before. The key to predictive safety management is to use whatever is deemed appropriate to enable the forward-looking approach.

Data mining. Data mining is a powerful way to find meaningful patterns, otherwise undetected, in large amounts of information. One author calls it a "process of finding interesting structure in data" through "induction-based learning" (Roiger, 2017, p. 42).

Monte Carlo simulation. Predicting the next accident can be challenging, given the range of factors and probabilities. A Monte Carlo simulation can help by presenting a range of outputs for given variables in a relatively short time period. The Project Management Institute explains the Monte Carlo simulation as

> an analysis technique where a computer model is iterated many times, with the input values chosen at random for each iteration driven by the input data, including probability distributions and probabilistic branches. Outputs are generated to represent the range of possible outcomes. (2017, p. 711)

Probabilistic Risk Analysis (PRA). This is a complex tool used to predict failure in intricate systems. Using a breakdown methodology similar to fault-tree analysis or work breakdown structures, PRA is based on the lowest component where data is available. Failure analysis for each component is combined with probability estimates to derive "the overall frequency of accidents for the system in question" (Bier, 1999, p. 704)

BEST PRACTICES/LESSONS LEARNED

Relative to the four components of SMS, the three previous tools do not address all of them. They are mostly related to the safety risk management and safety assurance categories. There are other opportunities in the safety policy and safety promotion components.

Communication tools. The ability to communicate effectively is one of the most important tools in the safety practitioner's toolbox. The safety professional cannot operate in a vacuum, and a healthy SMS requires active participation from the entire organization. A key example is the need for feedback, which requires communication. A good practice is to survey the needs of the organization, sometimes called the stakeholder analysis, and tailor communication to their needs and preferred methods.

Gap analysis. The gap analysis is another tool ranging in complexity, according to the need of the user. Its importance is recognized by ICAO, calling the gap analysis "central" to implementing the first phase of the SMS (2013, pp. 5–32). An extract of the ICAO example gap analysis checklist can be found in figure 2.4.

The basis of a gap analysis is to assist the organization in determining what the end result or need is. Once the need is defined, an analysis of the current

No.	*Aspect to be analysed or question to be answered*	*Answer*	*Status of implementation*
Component 1 — SAFETY POLICY AND OBJECTIVES			
Element 1.1 — Management commitment and responsibility			
1.1-1	Is there a safety policy in place? [5.3.7 to 5.3.15; 5.5.3]	☐ Yes ☐ No ☐ Partial	
1.1-2	Does the safety policy reflect senior management's commitment regarding safety management? [5.3.7 to 5.3.15]	☐ Yes ☐ No ☐ Partial	
1.1-3	Is the safety policy appropriate to the size, nature and complexity of the organization? [5.3.7 to 5.3.15]	☐ Yes ☐ No ☐ Partial	
1.1-4	Is the safety policy relevant to aviation safety? [5.3.7 to 5.3.15]	☐ Yes ☐ No ☐ Partial	

Figure 2.4 Extract of ICAO's Gap Analysis Checklist.

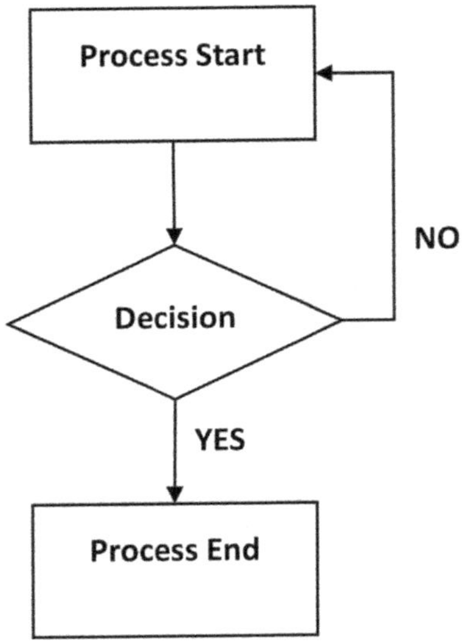

Figure 2.5 Flowchart Example.

situation is matched against that requirement. The delta between the two sides is the gap to bridge in order to meet the requirement.

Flow chart. Flowcharting is a process-mapping tool to help identify and visualize how a particular operation is conducted. Flowcharts are developed using standardized icons and they show various pathways, inputs, outputs, and decision points. An example of a flowchart is provided in figure 2.5.

Meetings. Well-planned and executed meetings provide a venue for crucial communication and feedback, both integral components of a healthy SMS. Regular meetings with key personnel across, as well as up and down the organization must take place in order to communicate critical information to be addressed within the system.

Process Decision Program Charts (PDPC). The PDPC is an extension of the root-cause analysis where the top event or condition is progressively broken into its component parts. Where the PDPC differs is in the addition of countermeasures. The countermeasures are developed to neutralize the bottom-level causes to stop the development of the top-level event (American Society for Quality, n.d.).

Information systems. Documentation and data collection are necessary enablers of the SMS. It is important to spend some time and resources on the organization's information system because of its foundational aspects. Project managers generically refer to this as a Project Management Information System (PMI, 2017), affirming this tool has broad applicability.

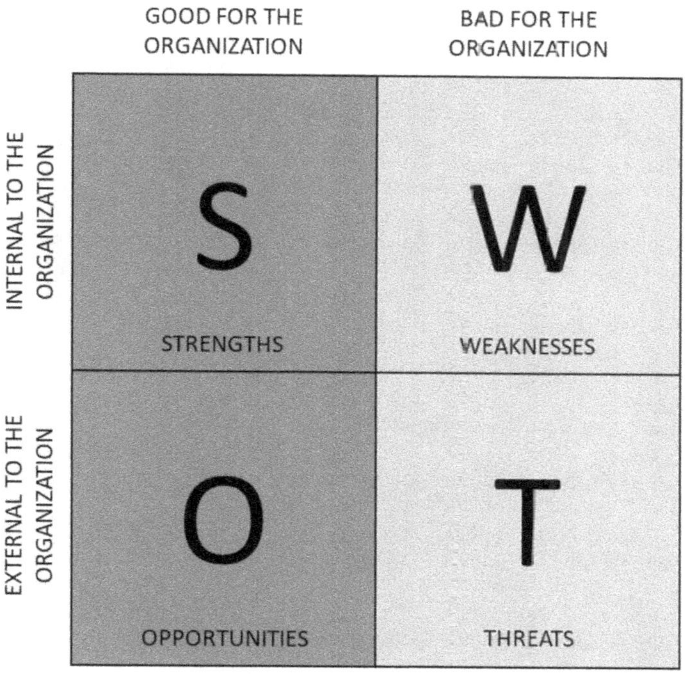

Figure 2.6 SWOT Diagram.

SWOT analysis. SWOT is an acronym defined as strengths, weaknesses, opportunities, threats. This tool provides a framework for identifying what the organization is doing well and any areas that may need improvement. See figure 2.6.

While common in leadership and business settings, the SWOT analysis can provide insight into the organization's safety programs and help support the SMS.

CONCLUSION

This chapter introduced several tools the safety practitioner can employ in establishing and maintaining the SMS. Admittedly, this is not an exhaustive list of tools. However, they are representative of the types of tools available to the safety practitioner. As a review, a consolidated listing of the tools explained in this chapter can be found in table 2.1.

Observation of experts in various field yields a common pattern. They all have "go-to" tools they like to use. However, not all tools are useful or

Table 2.1 Selected SMS Tools Matrix

	Program Categories		
Tool Name	*Reactive*	*Proactive*	*Predictive*
Accident Report Analysis	X		
Aviation Safety Action Program (ASAP)		X	
Bow-tie Diagram		X	
Brainstorming		X	X
Cause-and-effect Diagramming	X		
Checklists	X		
Check-sheet		X	
Communications Tools	X	X	X
Data Mining			X
Expert Judgment		X	X
Flow Charting	X	X	
Flight Operational Quality Assurance (FOQA)		X	x
Gap Analysis		X	
Human Factors Analysis and Classification System (HFACS)	X	x	
Internal Evaluation Program (IEP)		X	
Information systems	X	X	X
Inspections		X	
Interviews	X	X	X
Line Operations Safety Assurance (LOSA)		X	
Meetings	x	x	x
Mishap Investigations	X		
Mishap Plans	X		
Monte Carlo Simulations			X
Probabilistic Risk Analysis			X
Process Decision Program Charts		X	
Risk Matrix		X	
Root Cause Analysis	X		
Strength-Weakness-Opportunity-Threat (SWOT) Analysis		X	
Threat and Error Management		X	
Voluntary Disclosure Reporting Program (VDRP)		X	

Note: While it is possible to use the various tools in multiple situation, the major uses are identified with the upper-case X. Common secondary uses are identified with the lower-case x.

optimal in every instance. This is also true in the safety profession. Becoming familiar with common tools and keeping an eye out for new tools is an important part of being able to effectively implement and operate the SMS.

REFERENCES

American Society for Quality. (n.d.). *Process Decision Program Chart (PDPC).* Retrieved from http://asq.org.

American Society for Quality. (2017). *What is Root Cause Analysis (RCA)?* Retrieved from http://asq.org/learn-about-quality/root-cause-analysis/overview/overview.html.

Bier, V. M. (1999). Challenges to the acceptance of probabilistic risk analysis. *Risk Analysis, 19*(4), 703–710.

Donaldson, A., Borys, D., & Finch, C. F. (2013). Understanding safety management system applicability in community sport. *Safety Science, 60*(Supplement C), 95–104. doi:10.1016/j.ssci.2013.06.007.

Evans, J. (2017). Shimano undertaking investigation into Team Sky wheel failures at Tirren-Adriatico. *Cycling News*. Retrieved from http://www.cyclingnews.com.

Favarò, F. M., & Saleh, J. H. (2016). Toward risk assessment 2.0: Safety supervisory control and model-based hazard monitoring for risk-informed safety interventions. *Reliability Engineering & System Safety, 152*(Supplement C), 316–330. doi:10.1016/j.ress.2016.03.022.

Federal Aviation Administration. (2002). *Aviation Safety Action Program (ASAP) (AC 120-66B)*. Washington, DC: U.S. Department of Transportation. Retrieved from http://www.faa.gov.

Federal Aviation Administration. (2004). *Flight Operational Quality Assurance (AC 120-82)*. Washington, DC: U.S. Department of Transportation. Retrieved from http://www.faa.gov.

Federal Aviation Administration. (2006a). *Air Carrier Internal Evaluation Programs (AC 120-59A)*. Washington, DC: U.S. Department of Transportation. Retrieved from http://www.faa.gov.

Federal Aviation Administration. (2006b). *Line Operations Safety Audits (AC 120-90)*. Washington, DC: U.S. Department of Transportation. Retrieved from http://www.faa.gov.

Federal Aviation Administration. (2009). *Voluntary Disclosure Reporting Program (AC 00-58B)*. Washington, DC: U.S. Department of Transportation. Retrieved from http://www.faa.gov.

Federal Aviation Administration. (2015). *Safety Management Systems for Aviation Service Providers (AC 120-92B)*. Washington, DC: U.S. Department of Transportation. Retrieved from http://www.faa.gov.

Hakeem-Ur-Rehman, & Aslam, F. (2013). Implementation of quality improvement tools and techniques in conjunction with ISO 22000: A case study of dry roasted cashew nut processing plant. *Science International, 25*(3). Retrieved from http://go.galegroup.com.ezproxy.libproxy.db.erau.edu/ps/i.do?p=AONE&sw=w&u=embry&v=2.1&it=r&id=GALE%7CA361305595&sid=summon&asid=71bdab5ecef6c618fc7cf1070e337906.

International Civil Aviation Organization. (n.d.). *Accident Investigation Section (AIG): Documents*. Retrieved from https://www.icao.int/safety/airnavigation/AIG/Pages/Documents.aspx.

International Civil Aviation Organization. (2005). *Threat and Error Management (TEM) in Air Traffic Control* (Prelim ed.). Montreal, Canada: International Civil Aviation Organization. Retrieved from https://www.skybrary.aero.

International Civil Aviation Organization. (2013). *Safety Management Manual (SMM) (ICAO doc 9859)*. Montreal, Canada: International Civil Aviation Organization.

International Organization for Standardization. (2005). *Quality Management Systems – Fundamentals and Vocabulary* (3rd ed.). Geneva, Switzerland: International Organization for Standardization.

i-Sight.com. (2017). *The PEACE Model of Investigation Interviews.* Retrieved from http://www.i-sight.com.

iSixSigma. (n.d.). *The Cause and Effect (a.k.a. fishbone) Diagram.* Retrieved from https://www.isixsigma.com.

National Safety Council. (2017). *How to Conduct an Incident Investigation.* Retrieved from http://www.nsc.org/JSEWorkplaceDocuments/How-To-Conduct-An-Incident-Investigation.PDF.

National Transportation Safety Board. (n.d.). *The Investigative Process.* Retrieved from https://www.ntsb.gov/investigations/process/Pages/default.aspx.

OARS. (n.d.). *What Makes a Good River Guide Great?* Retrieved from https://www.oras.com/blog/makes-good/river/guide/great/.

Occupational Health and Safety Administration. (n.d.). *Incident Investigation.* Retrieved from https://www.osha.gov/dcsp/products/topics/incidentinvestigation/index.html.

PRO. (2017). *Statement: Re. Gianni Moscron/PRO 3-Spoke Wheel.* Retrieved from https://www.pro-bikegear.com/content/dam/probikegear/pressReleases/PRO%203-SPOKE%20WHEEL%20-%20STATEMENT.pdf.

Project Management Institute. (2017). *A Guide to the Project Management Body of Knowledge (PMBOK® guide)* (6th ed.). Newtown Square, Pennsylvania: Project Management Institute, Inc.

Reason, J. (1990). *Human Error.* New York: Cambridge University Press.

Red Bull. (2017). *Red Bull Rampage 2017.* Retrieved from http://www.redbull.com.

Roiger, R. J. (2017). *Data Mining: A Tutorial-based Primer* (2nd ed.). Boca Raton, FL: CRC Press.

SKYbrary. (2017). *Bow Tie Risk Management Methodology.* Retrieved from https://www.skybrary.aero/index.php/Bow_Tie_Risk_Management_Methodology.

U.S. Air Force. (2016). *Aviation Safety Investigations and Reports (AFMAN 91-223).* Retrieved from http://www.e-publishing.af.mil.

U.S. Army. (2015). *Accident Investigator's Handbook.* Retrieved from https://safety.army.mil.

U.S. Department of the Interior. (2014). *Interagency Aviation Mishap Response Guide and Checklist.* Retrieved from https://www.doi.gov/aviation/safety/iamrgc.

U.S. Navy. (2005). *Navy & Marine Corps Mishap and Safety Investigation, Reporting, and Record Keeping Manual (OPNAVINST 5102.1D).* Retrieved from http://www.public.navy.mil/NAVSAFECEN/Pages/instructions/OPNAVINST_5102-1.aspx.

Wiegmann, D. A., & Shappell, S. A. (2003). *A Human Error Approach to Accident Analysis.* Burlington, VT: Ashgate.

Chapter 3

Safety Management System Implementation Planning

David Thirtyacre

This chapter is intended to outline the process for SMS implementation. The concepts and processes ingrained within SMS are relatively straightforward, right up to the implementation phase. Turning SMS into practice can be a challenging endeavor and will differ in every application. Some organizations may have mature safety protocols in place that just need to be repackaged, while others are starting from scratch. As Stolzer and Goglia (2015) eloquently point out, "Expecting any organization to transition to a mature SMS within a short period of time is both unrealistic and self-defeating . . . . It is, and should be, a slow, methodical effort" (p. 218). Additionally, the lynchpin in SMS implementation is unwavering support from leadership, as the introduction of SMS should not begin without this support.

While SMS concepts apply to any organization, the aviation profession has embraced SMS, and many of the examples in this chapter will reference the aviation industry (see Appendix A for sample SMS outline for a small operation). Regardless of the industry, the foundational plan for implementation will *resemble* that of the aviation profession. This chapter will cover the planning required to define roles, assigned responsibilities, and establish accountability. The chapter will present theoretical and practical examples of SMS implementation in aviation and industry.

DEFINING AN IMPLEMENTATION PLAN

The implementation plan differs for every application of SMS and should be tailored with respect to each organization. Several published implementation plans are available for reference, and this section compares/contrasts their attributes. Each organization should customize their plan, but the procedures here are an excellent starting point. This chapter will highlight examples including Yu and Hunt (2004), the FAA (2015), and the ICAO (2013); although, other examples are available from organizations such as the American National Standards Institute (ANSI/AIHA Z10-2012), British Standards Institute (OHSAS 18001:2007), and the Occupational Safety and Health Administration (Voluntary Protection Program). While Yu and Hunt (2004) provide guidance toward a more general plan, the FAA (2015) plan is focused on compliance with AC 120-92B, and the ICAO (2013) plan is for implementation at the sovereign state level. Although the specifics of each plan will differ, they all introduce a phased approach to implementation, which defines a reasonable timeline and serves as guidance for SMS implementation. A common thread in all the examples is that of the four components of SMS: safety policy, safety risk management, safety assurance, and safety promotion. The key components also include a thorough understanding of the organization's current safety protocols and the desired objectives of the SMS.

YU AND HUNT

Yu and Hunt (2004) describe a systematic implementation plan based on total quality management (TQM) principles which consists of the following six phases:

Phase 1. Review of safety policy and safety plan.
Phase 2. Examine hazard identification and control plans.
Phase 3. Evaluate safety management practices.
Phase 4. Appraise incident investigation and emergency plans.
Phase 5. Analyze safety communication and documentation.
Phase 6. Review safety program evaluation and audits. (pp. 212–214)

Yu and Hunt (2004) further surmise that "the integration of TQM with SMS is perhaps easier to say that it is to do" (p. 214) and recommend four strategies for SMS implementation. These strategies, or overarching principles, should often be referenced during SMS implementation. They should be seen as the foundation of any SMS regardless of the selected implementation method or industry. The four strategies are as follows:

1. Organizational and cultural changes.
2. Full commitment and participation from staff.
3. Clarity of roles and expectations.
4. Long-term focus (pp. 214–215).

THE FEDERAL AVIATION ADMINISTRATION

The FAA identifies four broad areas for creating an SMS implementation plan. While these areas, listed below, are specific to a flying organization's compliance with FAA SMS regulations, they point out the importance of understanding the current safety policies and culture of the organization implementing the SMS. Note that the first two areas are similar to Yu and Hunt's (2004) Phases 1 and 3, highlighting the importance of understanding the current state of the organization.

- Mapping and analyzing your existing organization.
- Determining the extent to which your organization already complies with the requirements of 14 CFR part five.
- Developing a plan to comply with the requirements of part 5 with which your organization does not already comply.
- Submitting a plan to the FAA for approval (p. 47).

The FAA (2015) further breaks down the implementation plan into four levels and they highlight important ideas. Level 1 begins when the organization's leadership commits to providing necessary resources to support SMS implementation. Level 2 includes the implementation of safety risk management (SRM) and safety assurance (SA) and is referred to as the "reactive phase." Level 3 applies SRM and SA in a proactive or predictive manner, and Level 4 allows for continuous improvement.

INTERNATIONAL CIVIL AVIATION ORGANIZATION

The ICAO Safety Management Manual (SMM, 2013) provides a detailed implementation plan that contains many aspects applicable to all organizations. ICAO, similar to the FAA, proposes SMS implementation in a four-phase approach. These phases are broken down into significant detail and will be further discussed later in this chapter. The major contributions of each phase are listed here:

Phase 1: Assign responsibilities, perform a gap analysis, and develop the SMS implementation plan.

Phase 2: Implement essential safety processes, correct deficiencies, consolidate existing safety activities, and develop new safety processes.

Phase 3: Establish a risk management process to include collection and analysis of safety-related data.

Phase 4: Fully implement risk management and safety assurance through monitoring, feedback, and corrective action.

PREPARING TO PLAN

The precursor to successful SMS implementation is total commitment from leadership. Chen and Chen (2012) note the importance of leadership commitment, "Based on this concept, as policy maker, top managers are obligated to demonstrate their appreciation of SMS and commitment to its execution" (p. 177). The commitment may come from past safety losses, concern for the bottom line, or governmental regulations. Regardless of the motivation, the commitment must be genuine and complete, or the safety plan is destined to become a paper tiger—a safety program based on filing paperwork versus improving safety.

The main resistance to full commitment is often resource allocation. While leadership may agree that safety must be improved, if implementing SMS will negatively affect the bottom line or the report to shareholders, garnering support can be difficult. In the aviation industry, some businesses chose to invest in SMS while others did not. Federal regulations were required to motivate all aviation organizations to embrace SMS. This highlights the continued battle between the bottom line and safety.

The complexity of the organization and its mission will dictate the level of required planning. However, the plan, regardless of complexity, should always be mapped out with an overall flow of SMS implementation and started soon after deciding to incorporate SMS into the organization. These timeline products will serve as a guide to implementing SMS, by providing a blueprint for the process, establishing accountable timelines, and monitoring the implementation process. These steps may be adapted into checklists. The timelines should be thought of as part of a living document, anticipating numerous changes and additions as the process unfolds. There are several options to document the plan, and the FAA (2015) offers a sample, a portion of which is depicted in figure 3.1.

The FAA (2015) example identifies the task in enough detail to understand, sets an expected duration to complete the task, details when the task should begin and end, and provides a waterfall-type visual depiction of the timeline. As mentioned, this is a living document and should be routinely updated by the SMS implementation team. It is also an excellent tool to keep key leadership updated on implementation progress.

The sections that follow highlight critical areas of implementing the SMS plan. The depth and order of these areas will change, depending on the size

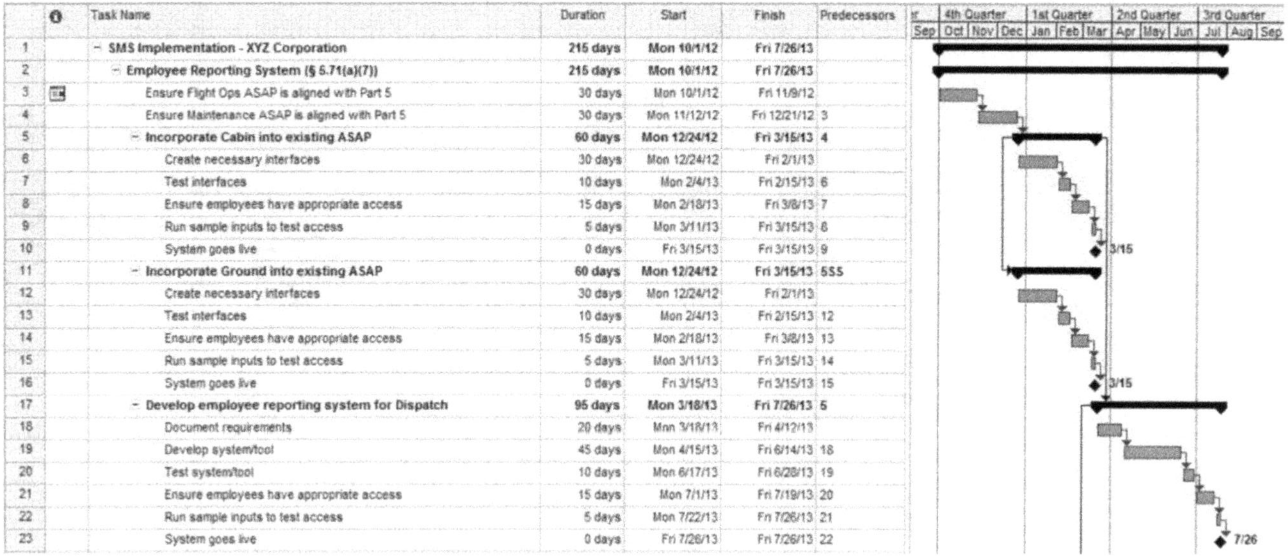

		Task Name	Duration	Start	Finish	Predecessors
1		− **SMS Implementation - XYZ Corporation**	**215 days**	**Mon 10/1/12**	**Fri 7/26/13**	
2		− **Employee Reporting System (§ 5.71(a)(7))**	**215 days**	**Mon 10/1/12**	**Fri 7/26/13**	
3		Ensure Flight Ops ASAP is aligned with Part 5	30 days	Mon 10/1/12	Fri 11/9/12	
4		Ensure Maintenance ASAP is aligned with Part 5	30 days	Mon 11/12/12	Fri 12/21/12	3
5		− **Incorporate Cabin into existing ASAP**	**60 days**	**Mon 12/24/12**	**Fri 3/15/13**	**4**
6		Create necessary interfaces	30 days	Mon 12/24/12	Fri 2/1/13	
7		Test interfaces	10 days	Mon 2/4/13	Fri 2/15/13	6
8		Ensure employees have appropriate access	15 days	Mon 2/18/13	Fri 3/8/13	7
9		Run sample inputs to test access	5 days	Mon 3/11/13	Fri 3/15/13	8
10		System goes live	0 days	Fri 3/15/13	Fri 3/15/13	9
11		− **Incorporate Ground into existing ASAP**	**60 days**	**Mon 12/24/12**	**Fri 3/15/13**	**5SS**
12		Create necessary interfaces	30 days	Mon 12/24/12	Fri 2/1/13	
13		Test interfaces	10 days	Mon 2/4/13	Fri 2/15/13	12
14		Ensure employees have appropriate access	15 days	Mon 2/18/13	Fri 3/8/13	13
15		Run sample inputs to test access	5 days	Mon 3/11/13	Fri 3/15/13	14
16		System goes live	0 days	Fri 3/15/13	Fri 3/15/13	15
17		− **Develop employee reporting system for Dispatch**	**95 days**	**Mon 3/18/13**	**Fri 7/26/13**	**5**
18		Document requirements	20 days	Mon 3/18/13	Fri 4/12/13	
19		Develop system/tool	45 days	Mon 4/15/13	Fri 6/14/13	18
20		Test system/tool	10 days	Mon 6/17/13	Fri 6/28/13	19
21		Ensure employees have appropriate access	15 days	Mon 7/1/13	Fri 7/19/13	20
22		Run sample inputs to test access	5 days	Mon 7/22/13	Fri 7/26/13	21
23		System goes live	0 days	Fri 7/26/13	Fri 7/26/13	22

Figure 3.1 Sample Work Breakout Schedule.

and scope of the specific SMS and goals; however, each of these areas is critical to effective SMS implementation.

MAPPING AND ANALYZING THE ORGANIZATION

A thorough understanding of the organization's departments, functions, management personnel, unions, and external inputs is the goal of mapping the organization. For complex organizations, this can be an arduous task. After the mapping is complete, the analysis begins and should, "describe and document what each department does, identify responsible and/or accountable personnel, and record who has the authority to modify its processes and procedures and accept risk for each department" (FAA, 2015, p. 48). The timeframe to accomplish mapping and analyzing the organization range from a few days for small businesses to a month or more for large organizations (FAA, 2015). The insight gained from mapping and analyzing the organization will begin to identify areas of interest for the gap analysis.

With the organization fully mapped, the safety gap analysis should begin to determine what safety policies and processes already exist, and more importantly, bring to light areas where the safety policy could be improved (ICAO, 2013). The safety compliance standard for the industry (e.g., FAA Part 5, OSHA) is an excellent starting point for the gap analysis. If an industry standard does not already exist, general standards can be created .

Most organizations already have some safety policies and procedures in place that will be identified during the gap analysis. The objective is to compare these policies, procedures, programs, systems, and activities to the SMS industry standards and determine where the deficiencies exist. Once the deficiencies are identified and the level of effort is determined, resources can be allocated, and personnel can be assigned to develop and implement solutions.

DEFINING ROLES AND ASSIGNING RESPONSIBILITIES

The SMS practitioner must balance the cost of SMS implementation with savings in life, equipment, hospital bills, lawsuits, and reputation. This is a delicate balance that must be considered early in implementation. Too little an investment and the program will never mature, and too much investment leads to inflated expectations and scrutiny. Another consideration is the scope of effort involved in implementation and the residual resources required to maintain the system after implementation. In any case, SMS calls for a relatively significant investment that will require research and planning to discover the correct mix for each organization and garner buy-in from the leadership element.

The required resources will not only include funding and personnel, but office space and equipment, access to personnel and processes at all levels of the organization, access to safety-related data, and authority to interact and communicate throughout the organization. For these reasons, it is highly recommended that the SMS is aligned to report directly to a high-ranking, accountable executive—often the president, CEO, or chief operating officer (COO) of the company. FAA AC 120-92B offers a flowchart to ensure the selected accountable executive is at the appropriate level in the organization (FAA, 2015). After the accountable executive is identified, the rest of the implementation team can be formed.

The composition of the implementation team should include representation from each *applicable* department. Determining which departments are relevant may take a few iterations. Often, seemingly insignificant departments are left off the team, but during the gap analysis, their input becomes important. While some positions on the team will be permanent and full-time, others will be temporary and part-time. In all cases, the team lead should be a person in a permanent position dedicated to SMS full-time. Once the team is established and an accountability chain-of-command is in place, the next immediate step is to develop the implementation plan.

With the senior accountable executive identified, the roles and responsibilities of the safety officials who will implement the program are determined. Regardless of the safety footprint in an organization, the responsibilities, duties, and authorities must be clearly delineated and formalized (Stolzer & Goglia, 2015). The authority delegated by the accountable executive will allow access to the organization's departments and personnel. While this requirement may seem ancillary to the larger SMS implementation, it is essential in order to properly articulate authorities and accountabilities and ensure they are known throughout the organization.

The size and makeup of the safety office is scalable but should always identify a single individual as the chief of safety. Ownership of the safety program is essential, whether as an additional duty or a full-time position, as it is the cornerstone of TQM. The chief of safety must be fully committed to SMS implementation and serve as the SMS champion. Lu and Yang (2010) concluded that genuine leadership influenced safety compliance and self-reporting significantly more than safety policy; the choice for chief of safety is critical. Other positions in the safety office can be divided along functional lines such as flight safety, ground safety, surgical safety, or explosive safety. Each safety representative should have extensive experience and knowledge in their respective areas. In larger organizations, there will be a need for positions such as data analysts, policy experts/advisors, accident/incident investigators, risk analysts, safety assurance personnel, and safety training/ education experts. Smaller organizations still require these functions, but they may be filled as part of additional duties or in multi-functional positions.

SAFETY POLICY, ACCOUNTABILITY, AND EMERGENCY RESPONSE

The formulation of safety policy is an important in SMS implementation. Many policies will be crafted throughout the process. However, producing a general policy that addresses the organization's overall guiding safety principles, establishes accountability, and highlights top management's commitment to the safety program aids in forming a solid foundation for implementation. The FAA (2015) provides a simple, one-page sample policy in Appendix 3 of AC 120-92B. While this sample policy is specifically written for air carrier application, it is an excellent example and can be easily adapted for use in any industry:

> The Executive Management of [Certificate Holder's name] recognizes that an effective SMS is vital to the success and longevity of the Company. Therefore, the executive management is committed to implementing and maintaining a fully functional SMS and to the continuous improvement of the level of safety throughout [Certificate Holder's name].
>
> - Executive management of [Certificate Holder's name] will establish specific safety-related objectives and will periodically publish and distribute to all employees those objectives and plans.
> - These safety objectives will be monitored, measured, and tracked to ensure overall corporate safety objectives are met. All employees and individuals in the company have the responsibility to perform their duties and activities in the safest practical manner.
> - [Certificate Holder's name] executive management is committed to providing the necessary financial, personnel, and other resources to establish and maintain a fully functional SMS.
> - [Certificate Holder's name] executive management is dedicated to establishing a confidential employee reporting system to report all hazards, accidents, incidents, and safety issues without fear of reprisal.
> - Activities involving intentional disregard for FAA regulations, company policies and procedures, illegal activities, and/or drugs or alcohol may be subject to disciplinary action.
> - As a component of the SMS, [Certificate Holder's name] executive management is committed to establishing, maintaining, and periodically exercising

an emergency response procedure and plan that provides for the safe transition from normal to emergency operations.

Executive management will convey this expectation to all employees through postings, intranet site, company newsletter, and any other means to ensure all employees are aware of the company's SMS, their duties and responsibilities, and our safety policy (FAA, 2015, Appendix 3).

Along with the overall safety policy, formalization of an emergency response plan (ERP) is required early in SMS implementation. Emphasizing the importance of an ERP, ICAO (2013) concludes that many organizations do not have effective plans in place to manage events during or following an emergency or crisis. How an organization fares in the aftermath of an accident or other emergency can depend on how well it handles the first few hours and days following a major safety event (Appendix 3).

Parts of the ERP may already exist in many organizations, whether formal or informal. However, the formal ERP should include procedures, responsibilities, coordination, and documentation, while identifying internal and external parties and their actions. ICAO (2013) Appendix 3 to chapter 5 identifies several specific considerations which an ERP should address. Depending on the SMS scope and application, many of these areas will expand into multipage documents and checklists and should be practiced through exercises:

- governing policies
- responding organizations
- notifications
- initial response
- additional assistance
- an emergency management center
- documentation and records
- accident site
- news media
- formal investigations
- family assistance
- post-occurrence review

SAFETY DATA COLLECTION AND ANALYSIS

The ability to collect safety data and properly analyze it is a cornerstone of SMS. While this process will continue throughout the use of SMS, the implementation plan should begin to discover how and what data are already collected, what needs to be collected, what methods should be used, and how it can be analyzed. The FAA (2015) identifies several steps to begin collecting and analyzing data early in SMS implementation. Below is a partial list from the FAA (2015) edited for applicability to all organizations.

1. Establish the context: Understand the safety performance objectives of the system, operations, or SMS.
2. Identify the objective of the analysis: Determine whether the analysis is for analyzing the safety performance of a system, of an operation, or the SMS itself.
3. Secure appropriate data: The data needed may be already on hand, or additional data-gathering may be needed, such as conducting a special audit with focus on a specific problem.
4. Select an appropriate data analysis method: For routine reporting, analysis may consist of tracking such things as dispatch reliability per month; system or part failure rates; crew utilization/duty time; and events such as minor incidents, diversions, and precautionary engine shutdowns.
5. Recommend: At this point, the person conducting the analysis may compare performance against relevant company safety objectives.
6. Document: Prepare reports and records in a format appropriate to your operation (p. 38).

An important part of data collection is creating an environment for honest, timely, self-reporting of safety-related issues. The foundational work for this process will be required during SMS implementation. To be effective, Leape (2002), in his research on safety reporting in the medical field, points out that the reporting system must be safe (anonymous and no threat of reprisal), simple (preferably one page), and worthwhile (taken seriously and analyzed). The FAA and ICAO echo this sentiment. The FAA (2015) specifically writes:

> In order to be effective, the organization needs to establish and maintain an environment in which employees feel comfortable to report hazards, issues, and concerns, as well as occurrences, incidents, etc., and propose safety solutions and improvements. The accountable executive and management team need to encourage employees to report safety issues and not fear reprisals from management. Policies that assure employees of fair treatment and clear standards of behavior are an essential part of the reporting process. (p. 36)

There may be other sources of data that already exists such as audit reports, employee reports, production efficiency, safety reports, and investigations. These sources can be an excellent starting point for analysis, since they are normally looked at in isolation instead of from an SMS-process point of view.

Data mining, statistical methods can be especially useful in cases where large amounts of seeming unrelated data are available.

RISK MANAGEMENT AND SAFETY ASSURANCE

The risk management process includes analyzing tasks and systems, identifying hazards, analyzing the risk, and applying controls to reduce risk (Stolzer & Goglia, 2015). While a comprehensive risk management process is an end goal of SMS, it can take several years to mature (FAA, 2015). During implementation, the safety experts should focus on designing the process of risk management and educating all levels of the organization.

Stolzer and Goglia (2015) propose sixteen components required in a process-based, safety risk management/safety assurance program. While the full description of each component is beyond the scope of this implementation chapter, the list sets forth a logical process flow ideal for early integration into a new SMS. Systems and tasks are broken down using this process, yielding mitigation and assessment strategies.

1. purpose
2. scope
3. interfaces
4. stakeholders
5. responsibility
6. authority
7. procedure description/flowchart
8. material/equipment/supplies
9. hazards
10. controls (production and protection)
11. process measures (production and protection)
12. targets (production and protection)
13. assessment (production and protection)
14. change management
15. qualifications required for participants in process
16. documentation/record-keeping (Stolzer & Goglia, 2015, p. 202)

In some cases, this is a quick, one-person process that can be accomplished in short order, while other cases require the coordination of several departments. During the implementation phase, the safety officer can identify (often from the gap analysis) a few tasks ripe for analysis and use them as an example during internal education and training.

EDUCATION AND TRAINING

Chen and Chen (2012) discuss the complications of turning a theoretical SMS into practice and note that, "It is thus essential to internalize the requirements of SMS into the organizational culture and the daily routines of individual employees, so that staff will know how to integrate the system with their own duties" (p. 177). The emphasis on "internalizing" the SMS is key to an effective SMS and can only be produced through strong policy backed by thorough, applicable education and training. The overall objective of education and training is to establish a long-term, just safety culture. The approach to education must be well planned, since it will most likely be the first exposure the practicing employee will have to SMS. While a majority of the training and education effort is spent while implementing SMS, it should be viewed as a continuing process with follow-on training and continuous improvement. Ironically, there is very little guidance for developing training and education from the FAA, ICAO, or OSHA.

The options for education and training include developing organic training material, fully outsourcing the training, or a using a mixture of both. The selected method will be based on resources and internal expertise. There are many companies and educational institutions specializing in SMS training with extensive, tailored options. A common solution is to outsource general SMS training (definitions and foundations, concepts, best practices, etc.) and create internal academics to cover company-specific policies and processes. Whatever the solution, a detailed plan, specifying the required training for each stage of SMS development, is required. Additionally, the training for the various levels of SMS oversight and position-specific training is required. The education and training plan should be integrated into the overall SMS implementation plan and tracked for quality and completion.

CONCLUSION

This chapter outlined several options and considerations for SMS implementation. While the concepts behind SMS are straightforward, incorporating SMS into a specific organization may be a complex and time-consuming endeavor. However, through careful planning and utilization of a phased approach to implementation, a fully functioning SMS will emerge. Scaling the implementation plan to fit the size, complexity, and mission of the organization, along with clear and unambiguous support from senior leadership, is required for success. This, along with continuous improvement and a willingness to accept change, can transform an organization into one that predicts and controls unacceptable risk instead of reacts to unanticipated accidents.

REFERENCES

Chen, S., & Chen, C. (2012). Scale development of safety management system evaluation for the airline industry. *Accident Analysis and Prevention, 47*, 177–181. doi:10.1016/j.aap.2012.01.012.

Federal Aviation Administration [FAA]. (2015). Safety management systems for aviation service providers. *Advisory Circular 120–92B.* Retrieved from http://www.faa.gov/documentLibrary/media/Advisory _Circular/AC_120-92B.pdf.

International civil aviation organization [ICAO]. (2013). *Safety Management Manual* (3rd ed.). Retrieved from https://www.icao.int/safety/SafetyManagement/Documents/Doc.9859.3rd%20Edition.alltext.en.pdf.

Leape, L. L. (2002). Reporting of adverse events. *The New England Journal of Medicine, 347*(20), 1633–1638. doi:10.1056/NEJMNEJMhpr011493.

Lu, C.-S., & Yang, C.-S. (2010). Safety leadership and safety behavior in container terminal operations. *Safety Science, 48*(2), 123–134. doi:10.1016/j.ssci.2009.05.003.

Stolzer, A. J., & Goglia, J. J. (2015). *Safety Management Systems in Aviation* (2nd ed.). Burlington, VT: Ashgate.

Yu, S., & Hunt, B. (2004). A fresh approach to safety management systems in Hong Kong. *The TQM Magazine, 16*(3), 210–215. doi:10.1108/09544780410532945.

Chapter 4

Organizing the Safety Management System

Michael Goldfein

Organizing the SMS for any business, regardless of the industry in which it resides, can be a daunting task. Once the company has decided to incorporate SMS in their organization in a way that will suit their requirements, there are a number of factors that must be considered. This chapter is designed to provide the specialist with the practical tools, not only to determine the size and shape of its SMS, but also to provide techniques for integration into the organization's daily routine. Every company, in various industries across the globe, has unique characteristics; therefore, it is important to develop a tailor-made program to fit the needs of the organization. The following seven phases or steps will provide a recipe for success.

The first step to creating an effective SMS is to calculate the breadth of the program. This prevents the SMS development team from getting too far into the process and then realizing it needs to be scaled back or built up more. The next step to creating a successful SMS is to designate the leader of the program, usually the president, CEO, or owner of the business, as the primary authority for decision-making power. Once established, the task of delegating authority to a prominent member of the organization is required, as the owner or CEO usually does not have the time or expertise to manage the SMS on a day-to-day basis. In some situations where an employee within the organization does not have the knowledge to take on this role, an outside source can be acquired or hired to fulfill this responsibility. This tends to be true for smaller companies that employ less than 50 personnel and do not possess enough safety know-how to fill the position from within.

Once the authority has been delegated to an SMS team leader, the task of providing resources begins. Resources are usually people scattered throughout the hierarchy of the organization, from top to bottom, so that every employee knows where to turn when a safety concern in the workplace arises. This step provides somewhat of a chain of command within the organization, in order to resolve issues at the lowest possible level. Once the resources have been chosen, the next step is to determine how to most optimally integrate the newly developed SMS. Once the integration phase is complete, it is time to consider purchasing whatever is needed to ensure a safe environment in which to operate. The purchasing phase would include materials, equipment, office supplies, and more to provide for the establishment of OSHA-approved and other approved items to be in compliance with minimum safety standards. Once the bulk of the SMS is developed and established, the emergency management plan should be created to prepare the organization for incidents and mishaps that may occur in the workplace or on the line. It is vitally important to define roles and responsibilities, generate response plans, and undertake the proper training to prevent disasters from occurring. The final phase, when shaping the SMS, is to establish the tools necessary to monitor and maintain the program throughout its lifespan. In this continuous examination period, the SMS is regularly scrutinized and modified to meet the needs of the organization. The following sections in this chapter will provide more detail to consider for each of the seven stages of SMS organization.

CALCULATING THE BREADTH OF THE PROGRAM

The primary consideration when determining the breadth of the program is the size of the organization, measured in terms of population. For example, a small, fixed-base operator (FBO) at an airport or an urgent care clinic in a small town would require a far less robust SMS than say a multimillion dollar company such as Southwest Airlines "employing over 55,000 people and flying over 4000 flights per day" (Southwest, 2017). In a large company like Southwest Airlines, a safety expert is hired specifically to fill the position of a safety manager and manage the day-to-day functions of the SMS. Under this safety manager, approximately 100 employees with varying safety expertise, from different divisions within the organization, will take on the role of a representative for each subdivision of the workforce. In essence, a chain of command will be established. When a safety concern is elevated, the safety manager does not need to be bombarded every day with small issues that can be resolved at the lower levels. In smaller organizations, one person may suffice as the designated safety manager, or if no expertise exists within the group, an outside specialist could be hired on full-time, or on a part-time, basis to direct the program. In addition to considering the number of personnel required to effectively manage the SMS, consideration must be given to the scope of the program. In other words, a gap analysis should be undertaken to compare the existing resources with the national or international standards and to outline the shortfalls within the system.

DELEGATING AUTHORITY

The delegation of authority is critical, especially among the larger corporations, to evenly distribute the obligations of managing the SMS. For the smaller business, there may be no need to delegate any responsibilities to lower echelons. However, for larger organizations to be more effective communicators, allocating some of the authority to the lower tiers is imperative to the success of the SMS. In a larger corporation, the safety manager should look at the organizational chart for the entire business and delegate approximately one member for every 500–1000 employees to become a part of the SMS staff. This allows the safety manager to distribute the staff evenly among the different divisions or subdivisions within the enterprise. Every group is different; thus, delegating authority could be based on population, diversity, expertise, or any partition that makes the most sense for their trade.

According to Naveh, Katz-Navon, and Stern (2011),

> To perform safely, the organization should develop a structure that supports safety activities. This structure includes key requirements such as establishing a clear manifesto of organizational safety policies and objectives. Second, the organization should delegate responsibilities for the development, implementation, and performance of safety, including designating a representative with the authority and responsibility for implementing and maintaining the safety management systems. In addition, the organization should arrange, establish, and maintain safety training programs that ensure all personnel are competent to carry out the safety requirements demanded by their duties and responsibilities. (p. 55)

Once these representatives have been chosen, a separate administrative diagram can be produced to portray the hierarchy among the SMS delegates. Following that, the safety manager can call for a meeting to explain the purpose of the SMS and describe the roles and obligations of the delegates within their respective sections. The effective training of those who are assigned safety responsibilities is critical to ensure employees are prepared and able to perform their duties optimally (Schaechtel, 1997). This will allow for each delegate to become the main point of contact for safety issues and concerns within their department. Only the concerns that require elevation up the chain of command would be presented to the safety manager, thereby allowing the manager to focus primarily on the big picture of safety and ensure SMS compliance within the organization. Once this suborganization has been established, the safety manager must provide the SMS team with the resources needed to function properly.

PROVIDING RESOURCES

This next phase provides the resources necessary to manage the SMS on a regular basis. The major sources of information available to the safety professional can be found online at the respective industry-specific websites at no cost to the end user. A majority of these websites are managed by governmental agencies responsible for the integration of the SMS into the workplace. For aviation-related companies, the FAA at https://www.faa.org and the ICAO at https://www.icao.int, will be the two primary locations for domestic and international flight safety, respectively. Another international source for Europe can be found at and the European Aviation Safety Agency (EASA) at https://www.easa.europa.eu. In addition, the National Transportation Safety Board (NTSB) site offers domestic accident and incident information and can be found at https://www.ntsb.gov. The National Aeronautics and Space Administration (NASA), also a United States government agency, is responsible for aeronautics and aerospace research as well as managing the civil space program. Their website can be found at https://www.nasa.gov. For non-aviation related resources, the primary website for OSHA is located at https://www.osha.gov, where a wealth of information can be discovered. Two primary locations to study workplace fatality and injury statistics can be found at the U. S. Department of Labor, https://www.dol.gov, and the Bureau of Labor Statistics, https://www.bls.gov. For international labor assistance, two agencies, the International Labor Organization (ILO), www.ilo.org, and the World Health Organization (WHO), www.who.int, are available. Now that SMS is becoming more of a buzzword in worldwide communities, simply typing "safety management systems" into the search criteria will bring up a plethora of assistance and safety guidelines.

In addition to free resources available on the worldwide web, organizations can take advantage of other personnel and programs made available through public enterprise. There are hundreds of companies whose primary purpose is to provide another set of eyes to the operation and offer best practices, after looking for hazards and any factors that might cause unsafe conditions or environments. Live safety instruction and computer-based training are just a couple examples of resources available to the employees within a company. After taking the advantage of the multitude of obtainable assets, the following phase would provide the safety managers with, for example, the FAA- or

OSHA-approved equipment and tools necessary to provide for their employees' health and safety in the workplace.

PURCHASING

Once the SMS skeleton has been created and the delegates have been given the chance to review the applicable minimum requirements as defined by the guidance pertinent to their specific organization, the time has come to purchase anything needed to sustain safe operations. Whether it is a simple, but potentially hazardous task, such as using a hammer and a nail, to the more complex and riskier task of operating heavy machinery, the correct protective equipment must be available for the job. As an example, these purchases would include approved equipment necessary to accomplish the task at hand while minimizing the risk. Once all the equipment, supplies, and references have been purchased and incorporated into the safety realm, the most difficult task is still yet ahead—the total integration of the SMS within the firm.

INTEGRATING

An integrated SMS is one in which every employee in the company understands their roles and responsibilities and can be held accountable for their own actions. According to Loebbake (2009), "Managers who involve themselves in the creation, implementation, rewards, and recognition of their SMS will create higher performing organizations" (p. 197). The manager must inform, educate, and demonstrate the different components of the SMS to the entire population within its jurisdiction and ensure each worker has the implements, tools, and gear necessary to accomplish their job safely. Everyone should feel comfortable knowing they have the right and the responsibility to tell someone else to stop a potentially hazardous activity and report up the safety channels, without fear of retribution for themselves or for their fellow employees. A fully integrated program allows for a safety-conscious workforce in a harmonious, agreed-upon atmosphere that promotes security, wellbeing, and protection at every corner of the operation. The success of the integration depends on the skills and the knowledge of everyone in the workforce and their ability to perform their job safely and competently. Stolzer and Goglia (2015) note,

> Responsibility for safety should be established at the appropriate levels throughout the organization. Usually this means that operations personnel should be tasked with safety. Good management dictates that if people are responsible for something they also should be held accountable. That means

> that job descriptions and performance reviews should reflect that responsibility. (p. 261)

Regardless of how well the SMS is integrated into the everyday operations of the organization, total accident and incident elimination is unrealistic. No perfect system exists, and although a zero mishap mentality is highly honorable, it is impractical. Based on this, an emergency management system must be created to minimize injury or loss of life if and when an incidence occurs.

EMERGENCY MANAGEMENT PLAN (EMP)

A logical and well-founded emergency management system needs to be created to prevent smaller occurrences from becoming larger tragedies. There are four main parts to the creation of an EMP:

1. Developing the plan.
2. Specifying the resources to implement the plan.
3. Training the employees to execute the plan.
4. Practicing the plan.

An effective EMP will be designed to protect people and emergency responders during an incident or accident.

MONITORING AND MAINTAINING THE SMS

Once all of the moving parts are fully integrated and running in the background, the simpler task of monitoring and maintaining the SMS ensues. The key to success in this final and perpetual phase is dependent on how well the integration and emergency management phases were employed. If each and every employee is truly committed to a safe operation and the emergency management team is prepared at a moment's notice to respond, the SMS is highly functional and ready to react and counter the impending threat. In this state, monitoring and maintaining the SMS should be stress-free. However, if the reverse is true, the safety manager and his delegates will be consistently looking for improvements to the system.

CONCLUSION

Organizing a SMS can be a time-consuming and challenging undertaking, depending on the size of the organization establishing it. The seven stages described in this chapter can be reduced or expanded, based on the needs of the organization. This is not intended to be a recipe for every association but, rather, an example for guidance when starting a SMS. As clearly stated earlier in the chapter, no two SMSs will match; rather, some will be simple and uncomplicated, while others may become very complex over time. John Loebbaka sums it up perfectly in an article discussing strategies and stakeholders when he states, "management ultimately bears responsibility for an organization's SMS and must foster stakeholder surveillance and involvement" (Loebbaka, 2009, p. 197).

REFERENCES

Loebbaka, J. K., & Lewis, A. (2009). Live or let die: Managing safety management system strategies and stakeholders. *Business Strategy Series, 10*(4), 193–198.

Naveh, E., Katz-Navon, T., & Stern, Z. (2011). The effect of safety management systems on continuous improvement of patient safety: The moderating role of safety climate and autonomy. *The Quality Management Journal, 18*(1), 54–67.

Schaechtel, D. (1997). How to build a safety management system. *Professional Safety, 42*(8), 22–24.

Southwest Airlines Company. (2017). Southwest corporate fact sheet. Retrieved from https://www.swamedia.com/pages/corporate-fact-sheet.

Stolzer, A., & Goglia, J. (2015). *Safety management systems in aviation*. Burlington, VT: Ashgate.

Chapter 5

Directing the Safety Management System

Roger C. Lee, Christopher Groom, Sarah Talley, and Marisa D. Aguiar

Over the past half century, commercial aviation has experienced dramatic growth in the global atmosphere. Associated with this growth has been a substancial reduction in accident rates (Boeing, 2015). With the experienced and anticipated growth in air traffic, the International Civil Aviation Organization (ICAO) is addressing global aviation safety by deployment of risk-based, safety management systems (ICAO, 2015). SMS has been referred to as the next step of improvement in safety (Müller, Wittmer, & Drax, 2014). Central to an effective SMS is a proactive, risk-based safety culture, shifting the emphasis of safety management away from a reactive approach focusing on compliances toward a more proactive approach. Stolzer, Halford, and Goglia (2011) described the need to establish a safety culture as part of the implementation of the SMS. Aviation is an industry consisting of a spectrum of stakeholders with diverse professional, national, geographical, and cultural backgrounds. Formulating a safety culture within this diversity is particularly challenging, but equally rewarding, if alignment is reached. An optimal safety culture cannot be achieved without good leadership within the organization. Leaders should take a top-down approach to safety, recognizing that the greatest influence on the success or failure of the SMS is their own level of commitment.

IMPETUS FOR EMBRACING SMS

SMS involves the people, processes, and equipment that must function in a synchronized manner to achieve an effective system. Having a technological solution without stakeholders embracing the SMS ethos will not deliver optimal results. Traditionally, ensuring safety has been often based on adherence to standard operating procedures (SOPs), the processes and procedures designed by the manufacturer, or the company management for the operator (flying, servicing, or maintenance) of aircraft. Giles (2013) highlights the dilemma faced by modern airline pilots in following SOPs with which they do not agree, understand, or feel adequately trained. McNeely (2012) points out that SMS looks deeper into contributing factors from all influencing sources.

UNSUCCESSFUL SAFETY MANAGEMENT

Edwards (2005) stated that reasons an organization's SMS may fail to deliver on its promise are due to:

- A lack of leadership and active involvement from senior managers
- A lack of commitment to the system providing the drive to make it work
- A lack of direction in the systems, where control is often devolved to middle manager levels
- A lack of understanding, by all but the specialists
- Middle management apathy
- The staff's belief that the initiative is just a fad

Most of the issues relate to the inability to direct a top-down SMS, rather than the inability to implement a particular technical methodology. Without a strong direction, SMS is no more than a set of tools having limited effectiveness in delivering a safe organization. Stolzer et al. (2015) stated the leading and directing role of SMS radiates from the management of an organization.

SAFETY CULTURE

A modern *SMS* is defined as, "a systematic approach to defining and achieving the desired safety performance" (FAA, 2013). This approach consists of four components: (1) safety policy, (2) safety risk management, (3) safety assurance, and (4) safety promotion (FAA, 2015). The approach is synonymous with how the members of an organization go about business. Deal and Kennedy (1982) describe this consistent approach as the *corporate culture* of an organization. When describing the approach that is taken to safety by all, the phase *safety culture* will be used.

DIRECTING A JUST CULTURE

When safety events occur, it is natural tendency to find individuals who are at fault and have them prosecuted. Mateou and Michaelides-Mateou (2012) provided a stark warning for those who breach the duty of care as proven by court, as they may be liable with serious legal consequences. Dekker (2012) states that offloading a failure onto a few individuals will not necessarily address the root causes of the problem nor prevent the mishap from happening again. Organizations must invest in the resources to identify *why* well-intended people can act mistakenly, without necessarily bad intentions, and without purposefully disregarding their duties or safety (Dekker, 2012). Within an effective SMS, organizations must spend the time and resources in investigating each safety event, identifying the root causes of each event, and applying mitigating measures to prevent them from recurring. From an SMS perspective, the more safety reports, the more accurate picture of the operational risks. The other perspective is that, if the information provided with only the best of intentions is used to identify breaches of regulations and primarily look for an offender, safety reporting will be suppressed. Dekker (2012) lists three central requirements to establishing a just safety culture within an organization:

i. Predictability in knowing who the party responsible is to draw the line between acceptable and unacceptable behavior.
ii. Role of domain expertise in drawing the line.
iii. Protection of safety data against interference.

Reason (1997) describes a just culture as an atmosphere of trust in which people are encouraged, even rewarded, for providing essential safety-related information while maintaining a clear boundary between acceptable and unacceptable behavior. People need to feel compelled to report safety infractions while knowing that the consequences of such reports are both fair and impartial. Management needs to firmly elucidate and distinguish between infractions and errors and ensure that the just culture model is followed for the SMS to be effective. For the SMS to be supported by a just culture, direction must be given throughout the organization to safeguard the three central requirements. Patankar, Brown, and Sabin (2012) describe four types of organizational cultures: secretive, blame, reporting, and just. Those with a just culture implement actions to ensure continuous improvement, rather than place blame on individuals diverting from the root cause. It is up to the middle managers to create and enforce an atmosphere that is conducive to the safety system.

WHO ARE THE STAKEHOLDERS?

Safety in commercial aviation cannot be delivered by a single entity. When describing the application of Crew Resource Management (CRM) in the aviation industry, Flin, O'Connor, and Mearns (2012) point to the importance of training each stakeholder on nontechnical skills to enhance performance such as leadership, situational awareness, decision-making, teamwork, and communication. Each stakeholder is the guardian of at least one particular safety barrier that prevents a hazard from becoming an accident. Aviation stakeholders who are on the front line and are critical in this process include the following:

- Pilots
- Engineers/Mechanics
- Ground service agents
- Air traffic controllers

Whenever there is an accident, the general public often focuses on the actions or inactions of frontline employees. Unfortunately, these individuals are not invincible and will make mistakes. Placing blame on these individuals will not make drastic changes to the likelihood of recurrence, as these individuals are at the end of the error chain (Wiegmann & Shappell, 2012). The rest of the error chain, the organization hierarchy, though not at the frontline, greatly influences safety. For example, management sitting in head office can formulate and reinforce error traps, or hazard conditions, such as lack of resources, schedule pressures on production, and unrealistic resource allocations (Dekker, 2006). To ensure critical stakeholders at all levels stay engaged and make safe decisions in operations, they must be identified and aligned with the SMS ethos. Various tools can be used to identify the stakeholders in the aviation system. One example is the 5M model—Man, Mission, Machine, Management, and Media (Stolzer et al., 2015). Perhaps one of the most common models is the SHELL model. A relatively simple example appears in table 5.1.

BEHAVIORAL CHANGES OF MANAGERS AND LEADERS

Building a culture rooted in safety requires input from all levels of management. A safety culture has to be collectively championed by employees at both the tactical and strategic level. This is imperative, as the climate of safety in any organization can fluctuate according to the prevailing state of

Table 5.1 Identification of Safety Stakeholders Using the SHELL Model

SHELL Model Element	*Element description*	*Stakeholders identified*
Software	"Software" the organizational inputs into flight crew through formal and informal training	Trainers / instructors Designer of Flight Training Flight training organizations Training platform (such as simulator) designers
Hardware	Aircraft Equipment	Designer of the aircraft Manufacturer of the aircraft
Environment	Physical environment such as weather and organizational environment	Meteorological experts Company senior management, CEO and senior management
Liveware	The physical and psychological condition of the pilot	Medical doctor Psychologists Family Members
Liveware-Liveware	People who interact with the pilot	Air traffic controllers Passengers Managers Engineers Ground / Ramp Agents

safety at any given time (Neal et al., 2000). As such, a culture of safety is the product of collaborations among safety management practices, interactions, conduct, and the outlook of top-level leaders expressed vicariously through shop level workers (Vinodkumar & Bhasi, 2009). The organization, as a whole, needs to empower the standard of safety with managers backing that state-of-mind with every tool in their arsenal.

When Lutchman, Maharaj, and Ghanem (2006) analyzed the research performed by Barratt (2000) on behavioral processes and building trust with employees, he observed that consistency in management's behavior inspires trust. Lutchman et al. (2006) further highlighted the importance of leaders ensuring their declared and actual priorities are consistent. It is destructive to the safety culture if safety leaders say safety is the priority but make decisions that undermine the importance of safety, such as denying appropriate financial resources.

Building on the concept of resilience, Stolzer et al. (2011) describe the link between resilience and actions in safety management. There is seldom an organization that is not under pressure from clients (internal or external), financiers, and others. The leaders who make the decisions must possess the communication skills to explain to clients and senior members of management why some unpopular decisions are being made. The reality is, if the explanation is not clear or not embraced by others, the safety leader will be seen as making unnecessary decisions and may ultimately be discredited, thus damaging the safety culture and management's credibility. Without trust, the SMS will not be effective. Safety leaders are required to stand firm in the beliefs that are stated in the company's safety policies, including open acceptance of safety reporting.

MEASURING SAFETY LEADERSHIP BEHAVIOR

Top-down strategic direction and implementation are important to the success of the SMS. Murcurio et al. (2002) highlighted various behavioral sets required by each level of safety leadership. Following are examples of measures of safety behavior at these levels:

- Top/mid-level managers: Measurement at this level includes personal behavior, safety activities, and statistical results. For example, following company safety rules, enforcing safety rules, arranging safety training, and monitoring workers' compensation costs.
- Supervisors: Measurement should include personal safety behavior and safety activities that they are able to control. For example, they need to make sure employees have safe materials and equipment, follow and enforce safety rules, conduct safety meetings, and maintain one-on-one contact with employees.
- Employees: Measurement usually includes personal behavior including complying with safety rules, participating in safety processes, and reporting injuries and hazards.

ORGANIZATIONAL DYNAMICS CHANGE MANAGEMENT

In terms of a strategy for the implementation of organizational change, Stolzer and Goglia (2015) described the Power and Interest grid, highlighting the importance of stakeholders and strategies to be adopted. A Power and Interest grid is a simple tool that assists managers in categorizing project stakeholders with increasing power and interest in the project. This tool allows one to focus on the key stakeholders who can make or break a project. In turn, tool assists in stakeholder prioritization. An example of a Power and Interest grid can be seen in figure 5.1.

Leaders must consider the key stakeholders as change is implemented and formulate a change plan in a structured manner. While going through the change process, resistance to change is to be an expected, natural, and healthy phenomenon (Coughlan, 1993). Coughlan stated that such resistance may be due to situational and personality factors but should still be taken seriously.

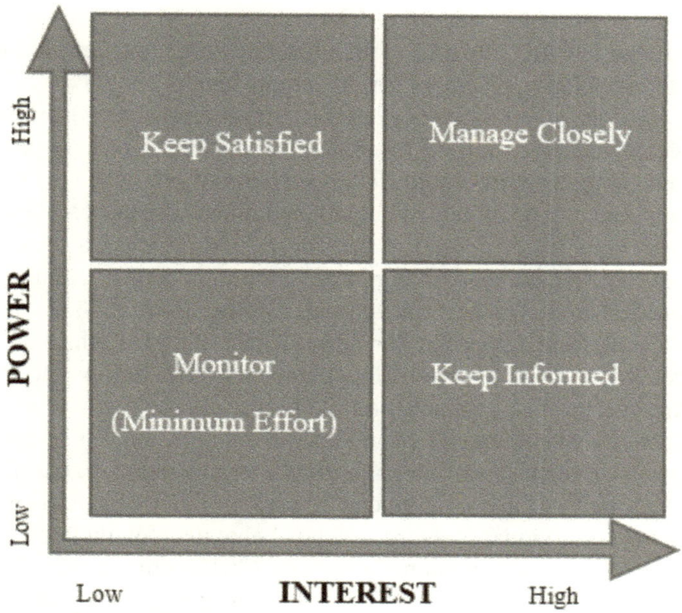

Figure 5.1 Power and Interest Grid.

Ignoring resistance or forcing change will result in increased resistance for the long term.

MOTIVATING THE STAKEHOLDERS

There are two common ways to motivate stakeholders—intrinsically or extrinsically. The following are a few recommended ways to optimize stakeholder motivation:

Self-protection. The need for safety is one of Maslow's Hierarchy of Needs (Maslow, 1954). Employees go to work expecting to go home safely every day. During SMS training for employees, it should be emphasized that the safety initiatives in place are there for their own protection and the protection of others.

Self-satisfaction and sense of achievement. Individuals will feel a sense of accomplishment when a job is done right (Wendover, 1995). On the practical side, doing it right the first time saves personal time, as well as the company. Personal pride in a job well done can be a lever a safety leader can pull in motivating personnel in the safety process.

Compliancy. Neal (2000) states that intrinsic motivation for safety behavior is largely based on compliance with safety rules and procedures. This concept is similar to the individuals' intrinsic value of not wanting to break the law. However, this intrinsic motivation does not consider the procedure or the rules as being appropriate for the circumstance.

Perception and psychological contract. Smith et al. (2013) discussed the link between perception and the psychological contract for employees. Safety leaders need to understand employee's perception of safety and the safety system. Perceptions can positively or negatively influence employees' buy-in into the safety system. If perceptions of the safety culture are not aligned with the employees' understanding of safety, motivation may be lost (Adam, 1965).

External rewards. In an educational setting, Rost (2013) stated that external rewards are important for unmotivated students who lack interest in college. Rost (2013) suggested that choices in goal setting and providing the appropriate level of challenge are likely to increase levels of motivation. Management may be tempted to provide cash rewards in terms of bonuses for staff to motivate them in meeting a particular organizational goal. Daalmans (2012) cautioned the use of financial reward to stimulate motivation as it risks seducing individuals toward solitary behaviors for their own benefit, rather than for the organization. Daalmans (2012) emphasizes the importance of promoting behaviors that ultimately result in safety, rather than just an increase in personal funds. On the topic of monetary reward, McCune et al. (2011) cautioned tactics that purely reward production targets, such as on-time delivery and cost efficiency, while the organization claims to have safety as a priority. Use of such production incentives may also encourage employees to cut corners (Patanka, 2012). Monetary reward for improvement of lagging safety indicators, such as injury rates and accident rates, were also questioned, as this has the tendency of encouraging employees not to report accurate data for the financial gain (Stolzer, Goglia, & Halford, 2011).

Recognition and praise. Instead of pointing out negative behaviors, a better approach is to highlight correct behaviors and reinforce them. There are various ways to achieve this (Perin, 2013):

- Complement the progress of employees more often than criticize it.
- Nominate a safety star of the month.
- Capture photographs of positive behavior and publicize them through safety magazines and other media.
- Invite the model safety ambassadors to give talks and share their experience and best practices.
- Invite the staff with positive safety behavior to write an article for company publications.

Irrelevant to monetary gain, individuals can be motivated by simple praise and recognition. This can be in the form of a formal letter of appreciation, a well-done message, a phone call, or a simple award, such as a pin or a badge. Leaders also need to objectively assess behaviors of staff to ensure the silent performers who follow all the necessary safety and quality practices are recognized and identified.

Career progression. Employees exhibiting positive safety behavior normally maintain other positive attributes such as diligence, attention to details, and safeguarding qualities. In terms of career progression, organizations are encouraged to consider safe behaviors as part of promotion criteria.

Safety Leadership and Communication

The term *leader* is used broadly here and refers to any person who needs to engage with stakeholders in helping them to embrace the SMS culture. As a leader (commander) of a B-52 bomber, Mosely (2008) wrote about the importance of communications from a CRM perspective; he concluded that communication is an integral part of a strong CRM. Safety leadership is about inspiring others so that the organization matures from the safety culture development.

For an aviation operator, employees can be scattered around the globe and work in different time zones. The only time staff visits the main office would be for hiring, induction, training, and job termination. Kanki et al. (2010) documented that the role of communication in a CRM context is to convey information, establish interpersonal/team relationships, establish predictable behavior and expectations, maintain attention to task and situational awareness, and use communication as a management tool. Kanki et al. (2010) goes on to mention four norms of communication in the safety leadership context. They are elaborated as follows:

I need to talk to you. Safety leaders very often have the need to communicate advice, instructions, and information to their stakeholders. Within the aviation environment, there is information flowing in all directions. Thus, safety leaders must communicate in a fashion that allows for information to be kept succinct and clear enough to be understood with little room of misinterpretation. Misunderstanding is the most common communication problem (Keshavarzi, 2011). In the aviation industry, with a wide range of cultural, national, professional, and experience differences, safety leaders must communicate in a way that is understandable by anyone, especially when delivering a critical safety message.

Safety leaders may come from technical backgrounds with little experience in the domains of public speaking or leadership. Olson (2009) found mumbling, poor grammar, monotonous tone of voice, talking too fast, profanity, and monopolizing the conversation all discouraged listeners. On the other hand, people will usually listen when speaking in a respectful tone, clearly and slowly; maintaining eye contact; and listening carefully to another's point of view without interrupting (Olson, 2009).

In terms of safety management, Maguire (2008) states that written communication provides a level of clarity and consistency that verbal communication lacks. Due to the criticality of the information being conveyed, safety leaders should consider having written backup of instructions or decisions supplementary to the verbal communication. Methods of communication should be commensurate with the criticality of the message.

I listen to you. Some companies operate a Read and Sign policy after issuing a safety notice to ensure employees understand the safety information being communicated to them. However, due to various human factors issues ranging from operational pressure to personal biases, there is no guarantee that the signature of the employee signifies they understand the information being presented. Therefore, safety leaders should not adopt a "fire and forget" attitude when issuing safety information. The feedback loop of communication must be assured. One of the ways to verify that the recipients have actually understood the information is to ask questions. Olson (2009) suggests safety leaders ask questions such as, "How is this going to change how you intended to do 'X'?" or "How long do you think it'll take you to accomplish this goal?"

On the feedback element, safety leaders are occupied by numerous, simultaneous demands from operational pressures to events in their personal lives. While they may give time to listen, they might not have achieved the true objective of listening, which is to understand. Imagine sharing a directive requiring the wearing of hard hats in a construction zone, for example. If the speaker simply said to, *make sure you wear your hats in the zone*, team members may interpret the word *hats* to mean corporate ball caps rather than safety protective devices.

Hoppe (2006) defined active listening as a person's willingness and ability to hear and understand. At its core, active listening is a state of mind that involves paying full and careful attention to the other person, avoiding premature judgment, reflecting understanding, clarifying information, summarizing, and sharing. By improving upon the skills and behaviors of active listening, leaders can become more effective listeners and, over time, improve their ability to lead. While safety leaders are presented with the information, they may be influenced by common perceptional deficiencies such as instincts, peer pressures, individual assumptions, group think, previous events, and inference. All are inherent hazards that may adversely affect the leader's ability to hear and make appropriate decisions.

I need you to talk to me. Safety leaders require the right information to make optimal decisions regarding the safety of operations. In the SMS, setting information flow can be formal or informal. It is crucial that employees feel comfortable to use the reporting system set in place by the organization, as a sound reporting culture is a key component to a safety system (Reason, 1997). Before employees can talk to the safety leaders, a level of trust must be built. The safety leaders must engage in a continuous campaign to reinforce the company's safety culture. One way to encourage the employees to talk to safety leaders is to conduct a survey into the subjective opinions employees have about various aspects of safety (Stolzer, 2011). Based on the data collected during this survey, safety leaders can use the information to understand the baseline view of safety within the organization and begin to develop mitigative solutions to assist in altering employee perceptions for the promotion of a positive safety culture.

Taylor (2010) mentioned that public pressures, regulatory interventions, and national cultures can all affect the safety beliefs of an organization. The cultural norms of the various individuals within the organization should be compensated for by the organizational culture. In a sense, when an employee enters the door of an organization, they are entering another country with a unique set of values, beliefs, and behaviors founded on safety culture. Zhohar (2011) calls this concept *safety citizenship.*

I expect you to talk to me. If a situation occurs when safety could be compromised within the organization, safety leaders must be informed. Routine in safety management are audits, investigation reports, communication of findings, and appropriate mitigating actions to address root causes of accidents and incidents. While audits and investigations are often required components of an organization's safety and quality systems, they do not always catch everything. A risk or hazard may go unnoticed. Thus, safety leaders must have an open line of communication with employees, where employees feel safe and protected enough to come to the safety manager with safety concerns they have observed on the line.

Getting employees to express safety concerns while growing a culture can be difficult at best to accomplish for managers and nearly impossible in certain climates depending on social barriers. To convey information, leaders need to use psychological tools to instill concepts in the employees. By expressing safety through a medium that is well understood by all and both emotionally and intuitively connective, safety communication can reach broader audiences and solidify SMS concepts faster. Alternatively, managers need to seek out ways to deliver their safety message to the largest audience with the most impact. This can be challenging to accomplish due to diversity within the organization but is possible with some finesse and assistance from others in leadership.

Communication Constraints

There are walls to communication all managers must overcome. Perhaps someone is soft spoken or too loud depending upon what is appropriate for a specific setting. By communicating ineffectively, even the most powerful of messages can fall on deaf ears. SMS professionals must be cognizant of their internal limitations and modify their delivery platform to reach the broadest of audiences, while finding solutions to their own personal communication constraints.

PROMOTING SAFETY

Incentive Programs

Managers must err on the side of caution when providing incentives that reward people for doing the right thing. This is especially true if the reward is monetary or gives folks time off from work, as it may encourage false reporting of information. According to a 2010 survey conducted by the Government Accountability Office, 75 percent of manufacturers in the United States had safety incentive programs that may be affecting workers' reporting of injuries and illnesses (Flynn & Short, 2012). This survey was driven by a concern that safety incentive programs may actually be encouraging non-reporting; after careful analysis, OSHA issued a memo supporting safety incentive programs, when done right (Flynn & Short, 2012). It is critical for safety managers to understand that the organization's safety climate will be tied to a carefully balanced safety incentives program. The National Safety Council recommends utilizing the following categories to measure and recognize leading indicators safe work, proactive safety, safety leadership, and safety engagement (Flynn & Short, 2012).

Category I: Safe Work. Employers should incentivize behaviors that positively influence safety. For example, employees could be rewarded for participation in a safety meeting, supervision-recognized safe act, zero incidents for the month, or perfect attendance—just to name a few.

Category II: Proactive Safety. It is the duty of upper-level management to echo the mantra that safety is the responsibility of everyone within the organization. Thus, employers should incentivize behaviors indicative of proactive safety. Examples could include the reporting of a near miss, reporting unsafe conditions, completing safety training, or participating in a safety audit.

Category III: Safety Leadership. The safety climate of an organization could be bolstered by promoting and rewarding exemplifications of safety leadership. Some of these behaviors include leading a safety meeting, participating in a root cause investigation, and supervisor-observed demonstrations of safety leadership, such as recognizing a fellow employee for a safe act.

Category IV: Safety Engagement. Employees should be rewarded, or at least recognized, when they go above and beyond for safety. This could be demonstrated through the behaviors the employees display daily while in the working environment but could also include bring safety suggestions to management. Safety suggestions stemming from employees should always be seriously considered by the safety manager, as the employees could provide the greatest insight into overlooked hazards.

LEADERS, MANAGERS, AND CHANGE OVERSIGHT

Safety process enforcers can be leaders, managers, or both. Leaders show methods and procedures by acting upon them in daily operations for others to mimic and follow. Managers spend their time subduing workers into a schedule of habits based on compliance, without much actual first-hand effort made into understanding or participating in objectives. The goal is to build small teams of leaders that can influence their inner circle of members to gain the most impact, rather than have a large-scale group of managers loosely enforcing principles.

Interactive Leadership

Managers can bolster employee perception about the changes brought about through implementing a SMS by providing mentorship and positively interacting with employees. The goal is to make the presences of the safety manager known within the organization. Building a rapport with employees may aid in improving the safety culture of the organization. The safety manager's presence will demonstrate the organization's commitment to safety culture and reinforce that safety is the responsibility of all. Safety should be garnered and built into the bedrock of a corporation through the *show-me-tell-me* method of teaching. Within this method, feedback and real-time dialogue exist between mentors and subordinates. This can only be accomplished when safety managers are actively making their presence known, while simultaneously building trust among employees.

Influential Leadership

There are three major leadership styles: transactional, transformational, and servant-based. In transactional leadership, leaders will enter into exchange relationships with employees in order to attain mutually acceptable goals (McQueen, 2013). Workers will reap both tangible and intangible benefits from their superiors, and supervisors receive the best quality work that they can get. The second type of leadership, transformational, occurs when leaders inspire followers to change expectations, perceptions, and motivations in pursuit of common goals (Lievens, Van Geit, & Coetsier, 1997). The result is a win-win situation where safety system professionals gain insight from their interactions with employees, resulting in vested interests on both sides.

The third primary leadership style is servant-based leadership. In servant-based leadership an emphasis on service motivation, as demonstrated by empowering and developing people with empathy and humility, differentiates this style from other leadership frameworks (Mittal, 2012). This style of leadership can be a highly effective method for building buy-in and often results in the most rewards reaped by both the superior and the subordinate. Good leaders tend to blend a balance of the three leadership styles and shift them appropriately based on the situation. These are fluid and dynamic methods that can be adapted based upon the situation at hand.

CREATING A CLIMATE OF SAFETY

The climate of safety is built upon employing multiple methods of instruction and using a plethora of tools that assist leadership in moving and sustaining the message of preparedness and risk aversion. Tools of this sort promote a climate of safety by holding people accountable for their actions and further breaking down the rationale behind the way decisions are made. By following procedures and guidance in conjunction with these tools and others, leaders can create a culture based on a safety mindset.

One tool in the leadership toolbox is the root cause analysis. Root cause analysis is a problem-solving approach that seeks out the root causes of problems rather than focus on causal factors that are implicated in but not necessary to the problem's occurrence (RCA, 2015). This tool zeros in on the originating point of an issue rather than on the symptoms. Often, when safety managers or designated investigators analyze events, they look at trends and outputs and disregard the very important point of figuring out exactly where problems originated. In the Japanese workplace, the *is sho ni* (together as one) culture, demonstrated by Toyota is a prime example of a collectivist safety culture where responsibility for safety is dispersed across the organization. Toyota is essentially the founder of the root cause problem dissection process. A great deal can be learned from their model. Root cause analysis allows for mitigation of hazards at the ground level.

Safety Culture and the Safety Climate

Leaders are the builders of the SMS social pyramid in an organization. A properly aligned safety culture is a critical component in building an organization that demonstrates, enforces, and displays safe attitudes and mindsets. Culture "heavily influences an individual's behavior by setting group norms" (Sharman, 2016, p. 23) and is at the very core of a safety system, as people emulate what they see and hear around them on a daily basis. Culture can be written into existence and is also built through beliefs of the people that make up an organization. To reinforce a specific cultural mindset, "safety audits have the objective of assuring that the necessary constituents for safety have been implemented and institutionalized" (Berglund, 2016, p. 7).

In the work center, safety climate perceptions should focus on the nature of relationships among safety policies, procedures, and practices (Zohar, 2010). Tools such as the Climate Assessment Survey allow workers the opportunity to express their opinions and beliefs for the purpose of providing managers with insight into the overall health of the organization's safety climate.

Safety Culture as an Attitude

In SMS, it is has become apparent that a safety culture, or lack thereof, can have a significant impact on a company's safety performance (Shannon, Mayr, & Haines, 1997). By adjusting attitudes, encouraging positivity, and heightening the focus on safety, these postive changes to the organization's safety culture will resonate throughout the organization. Effective safety cultures need to be built on the deliberate and intentional movement of positive thoughts and ideas if they are to succeed and stand the test of time. A common misconception is that safety exists as the absence of injury. However, safety is not about the absence of something, such as incidents or deviations, rather it is about the presence of something (Solomon, 2015). In some organizations, particularly those with more toxic attributes, *name, blame, shame, and retrain* is a widespread practice where leaders attack people verbally over mistakes. This approach fails to address underlying causes of incidents (Solomon, 2015). Research has shown that the safety attitude, motivation, and safety knowledge are the best predictors of safety behavior at workplaces (Mohammadfam, 2017). Thus, in order to promote a healthy safety culture and an overall effective SMS, leaders must maintain a positive attitude, and that same attitude will eventually infiltrate all levels of the organization.

Requirement of Safety versus Necessity of Safety

It is imperative for managers to distinguish between the requirements of safety and the necessity of safety in the workplace. Healthy organizations have safety requirements in place via policies and procedures designed to protect both the worker and the company from mishaps or legal ramifications. All organizations must manage risks to the health and safety of employees for the purposes of protection and legislation (Hill, 2009). Legislation comes in the form of OSHA standards, ICAO mandated practices, or a plethora of other sources that impact and shape the way in which safety is expressed or enforced in aviation. "Most organizations follow the Civil Aviation Authority guidelines and/or directly ICAO SMS framework which shows that best industry practices and regulatory parameters are the most commonly used" (Müller et al., 2014, p. 86). Guidance and models from this source are typically cutting-edge implementations for safety that are updated at a reasonably efficient pace. Particularly, critical safety mishap concerns are expressed and published as soon as possible to ensure that organizations are doing things safely and by the book. Managers need to instill in subordinates that safety is more than just a requirement—it is a necessity. For example, in the construction business the hard hat is a company's most visible commitment to safety. If all the employees are wearing their hard hats on a job site, it can be assumed that company has a good safety program (Smith, 1994).

Safety Assumptions

In any organization, leaders responsible for implementing and adhering to standards make certain assumptions about safety. There are two major categories of behaviors being considered: (1) unintentional errors such as slips, lapses, or mistakes and (2) the noncompliance with work rules and procedures (Reason, 1997). The first category is something that happens and is difficult to control. Training, policies, procedures, and other tools are used to combat these errors. With the second category, it is entirely possible to modify the behavior of individuals through training. Moreover, deviations from standard operating procedures must always be addressed. This can only work if assumptions are taken out of the equation, and everyone is held to the same standard of behavior.

Safety Expectations

Lead staff members need to champion safety and regard it as the expectation, not the exception. Safety is something to be expected and anticipated by all within an organization. According to Thatcher (2003), management can expect the employee's expectations of safety to include immunity from hazards; protection from unsafe equipment; and personal security in the workplace. Maintenance of an effective SMS will allow the organization to meet employees' safety expectations, although it is still up to the worker to have good situational awareness, perspicacity, and alertness on the job. It is the responsibility of management to confirm that employees are focused and motivated, and following standard operating procedures. Expectations must be clear and concise for safety to work, with zero tolerance for misbehavior.

CONCLUSIONS

In this chapter key concepts that leaders and managers must both embody and consider to effectively direct their SMS were reviewed. It is up to leadership to set the tone of safety within the organization, as the organization is only as good as its leadership. The success of the SMS requires the unwavering support of top management, which must be constantly fed through to the front line by key figures within the organization (Cacciabue, 2015). By channeling the SMS vision statement from the top down, leaders can help ensure employees at all levels are committed to upholding the safety standards of the organization.

As demonstrated in this chapter, directing the SMS is no easy task and requires extensive consideration into many facets to optimize safety. Usually, simple compliance is not enough to deliver safe performance. Safety managers are tasked with safeguarding the core values of the SMS by maintaining a just culture, promoting open and effective communication, and ingraining safety as a priority within all levels of the organization. Safety is the responsibility of everyone, and this chapter has provided the tools to effectively work with stakeholders on developing strategies based on organizational theories.

REFERENCES

Berglund, J. (2016). *Why Safety Cultures Degenerate: And How To Revive Them* (1st ed.). London, England: Routledge.

Boeing. (2015). Statistical summary of commercial jet airplane accidents. *Worldwide Operations: 1959 – 2014*. Retrieved from http://www.boeing.com/resources/boeingdotcom/company/about_bca/pdf/statsum.pdf.

Cacciabue, P. C., Cassani, M., Licata, V., Oddone, I., & Ottomaniello, A. (2015). A practical approach to assess risk in aviation domains for safety management systems. *Cognition, Technology & Work, 17*(2), 249–267.

Daalmans, J. (2012). *Human Behavior in Hazardous Situations: Best Practice Safety Management in the Chemical and Process Industries*. Saint Louis, MO: Butterworth-Heinemann.

Deal, T., & Kennedy, A. (1982). *Corporate Cultures: The Rites and Rituals of Corporate Life*. Reading, MA: Addison-Wesley.

Dekker, S. (2006). *The Field Guide to Understanding Human Error*. Burlington, VT: Ashgate.

Dekker, S. (2012). *Just Culture: Balancing Safety and Accountability* (2nd ed.). Burlington, VT: Ashgate.

Edwards, C. (2005). Shell aircraft international experience. *Transport Canada*. Retrieved from http://search.proquest.com.ezproxy.libproxy.db.erau.edu/docview/214604905?pq-origsite=summon.

Federal Aviation Administration. (2015). Safety management systems for aviation service providers. *AC No. 120-92b*. Retrieved from https://www.faa.gov/documentLibrary/media/Advisory_Circular/AC_120-92B.pdf.

Flin, R., O'Connor, P., & Mearns, K. (2002). Crew resource management: Improving teamwork in high reliability industries. *Team Performance Management, 8*(3), 68–78.

Flynn, R., & Short, S. (2012). Designing a safety incentive program. *National Safety Council*. Retrieved from https://www.safetyandhealthmagazine.com/articles/designing-a-safety-incentive-program-2.

Giles, C. N., & Giles, C. N. (2013). Modern airline pilots' quandary: Standard operating procedures–to comply or not to comply. *Journal of Aviation Technology and Engineering, 2*(2), 1.

Hill, N. (2009). Air necessities. *The Safety & Health Practitioner, 27*(6), 38.

Hoppe, M. H. (2006). *Active Listening: Improve Your Ability to Listen & Lead.* Greensboro, NC: Pfeiffer.

ICAO. (2015). Safety Report 2015 Edition. Retrieved from http://www.icao.int/safety/Documents/ICAO_ Safety_Report_2015_Web.pdf.

Kanki, B. G., Helmreich, R. L., & Anca, J. (Eds.). (2010). *Crew Resource Management* (2nd ed.). Burlington, MA: Academic Press.

Keshavarzi, Z. (2011). Misleading communication vs. effective aviation management. *Intelligent Information Management, 3*(6), 240–243.

Lievens, F., Van Geit, P., & Coetsier, P. (1997). Identification of transformational leadership qualities: An examination of potential biases. *European Journal of Work and Organizational Psychology, 6*(4), 415–430.

Lutchman, C., Maharaj, R., & Ghanem, W. (2006). *Safety Management: A Comprehensive Approach to Developing a Sustainable System* (1st ed.). Boca Raton: CRC Press.

Maguire, R. (2008). *Safety Cases and Safety Reports: Meaning, Motivation and Management.* Burlington, VT: Ashgate.

Mateou, A. C., & Michaelides-Mateou, S. D. (2012). *Flying in the Face of Criminalization.* Burlington, VT: Ashgate.

McCune, D., Lewis, C., & Arendt, D. (2011). Safety culture in your safety management system. In A. J. Stolzer, J. J. Goglia, & C. D. Halford (Eds.), *Implementation of Safety Management Systems in Aviation* (pp. 135–160). Burlington, VT: Ashgate.

McNeely, S. C. (2010). Lessons learned from toyota-2010-QMS vs SMS. *Federal Aviation Administration.* Retrieved from https://www.faa.gov/about/initiatives/sms/explained/quality_and_safety_management/media/QMS%20vs%20SMS-Lessons%20Learned%20from%20Toyota.pdf.

McQueen, M. R. (2013). The effects of Kurt Lewin's three major leadership styles on followers: A follower's perspective. *ProQuest Dissertations Publishing 3573682.* Retrieved from https://search-proquest-com.ezproxy.libproxy.db.erau.edu/docview/1437649356?clientId=17916&https://proquest.umi.com.ezproxy.libproxy.db.erau.edu/login?COPT=REJTPUcyODcrM2IwZiZJTlQ9MCZWRVI9Mg==&pq-origsite=summon.

Mittal, R., & Dorfman, P. W. (2012). Servant leadership across cultures. *Journal of World Business, 47*(4), 555.

Mohammadfam, I., Ghasemi, F., Kalatpour, O., & Moghimbeigi, A. (2017). Constructing a Bayesian network model for improving safety behavior of employees at workplaces. *Applied Ergonomics, 58*, 35–47.

Mosely, D. (2008). The role of communication in CRM. *Flying Safety, 64*(10), 6–7.

Müller, R., Wittmer, A., & Drax, C. (2014). *Aviation Risk and Safety Management: Methods and Applications in Aviation Organizations.* New York, NY: Springer.

Neal, A., Griffin, M. A., & Hart, P. M. (2000). The impact of organizational climate on safety climate and individual behavior. *Safety Science, 34*, 99–109.

Olson, J. (2007). Effective communication. *Surface Fabrication, 13*, 18.

Patankar, M. S., Brown, J. P., & Sabin, E. J. (2012). *Safety Culture: Building and Sustaining a Cultural Change in Aviation and Healthcare.* Burlington, VT: Ashgate.

Perin, M. (2013). The power of recognition and praise. *Officer.Com.* Retrieved from http://search.proquest.com.ezproxy.libproxy.db.erau.edu/docview/1423743663?accountid=27203.

Reason, J. T. (1997). *Managing the Risks of Organizational Accidents.* Burlington, VT: Ashgate.

Rost, M. (2013). *Active Listening*. Harlow, England: Routledge.

Shannon, H. S., Mayr, J., & Haines, T. (1997). Overview of the relationship between organizational and workplace factors and injury rates. *Safety Science, 26*(3), 201–217.

Sharman, A. (2016). *From Accidents to Zero: A Practical Guide to Improving Your Workplace Safety Culture* (2nd ed.). Boca Raton, FL: Taylor and Francis.

Smith, B. E. (2013). Prospective safety analysis and the complex aviation system. *NASA*. Retrieved from http://ntrs.nasa.gov/archive/nasa/casi.ntrs.nasa.gov/20140010408.pdf.

Smith, S. L. (1994). *Hard Hats: The Unsung Safety Necessity*. Cleveland, OH: Penton Media, Inc.

Solomon, B. (2015). Developing a robust safety culture. *Professional Safety, 60*(8), 50.

Stolzer, A. J., & Goglia, J. J. (2015). *Safety Management Systems in Aviation.* Burlington, VT: Ashgate.

Stolzer, A. J., Halford, C. D., & Goglia, J. J. (2011). *Implementing Safety Management Systems in Aviation*. Burlington, VT: Ashgate.

Thatcher, J. J. (2003). *Delivering on Safety Expectations.* Cleveland, OH: Penton Media, Inc.

Vinodkumar, M. N., & Bhasi, M. (2009). Safety climate factors and its relationship with accidents and personal attributes in the chemical industry. *Safety Science, 47*, 659–667.

Wiegmann, D. A., & Shappell, S. A. (2012). *A Human Error Approach to Aviation Accident Analysis: The Human Factors Analysis and Classification System.* Burlington, VT: Ashgate.

Zohar, D. (2010). Thirty years of safety climate research: Reflections and future directions. *Accident Analysis and Prevention, 42*(5), 1517–1522.

Chapter 6

Safety Performance Management

Rachelle L. Strong

The purpose of performance management is to measure the effectiveness of risk controls that have been implemented within a safety system (SMS; Stolzer & Goglia, 2015). According to the ICAO, performance management functions as safety assurance within an SMS. After defining the organizational objectives, progress is measured. Based on this, objectives and adjustments to performance are subsequently made. In safety management, these organizational objectives are referred to as "safety performance indicators" (SPIs), which are assigned labels, target values, and alert values. Some examples include cost and benefit targets for safety-related items, compliance with regulatory requirements, and public or customer expectations (ICAO, 2013).

SAFETY PERFORMANCE INDICATORS

Consider a situation where an SMS has been implemented, such as at a U.S. commercial airline. Fully implementing the SMS framework can take years, depending on the complexity and maturity of the organization; however, performance management begins at the same place every time—defining measurable performance outcomes to determine if the SMS is operating as it was designed, typically exceeding regulatory requirements (ICAO, 2013). Measuring SPIs helps monitor existing, documented risks, detect developing risks, and put corrective action plans in place. Performance management functions as a means of aiding continuous improvement for an SMS within an organization.

COMMERCIAL AIR CARRIER REGULATIONS

Considering the commercial airline example, in the United States, the regulatory agency is the FAA. The FAA released 14 CFR Part 5 which requires Part 121 air carriers to have a qualified SMS in place by March of 2018, which must be submitted to and approved by the FAA administrator. In terms of regulatory definition, safety performance means, "realized or actual safety accomplishment relative to the organization's safety objectives." In accordance with the ICAO SMS Manual (2013), SPIs, and the management thereof, can serve as documentation regarding their SMS effectiveness, which also function as evidence of regulatory compliance. In other words, for a commercial airline to meet FAA requirements, they need an established set of criteria to measure their SMS effectiveness regarding cost and benefit targets, risk reduction targets, compliance targets, and customer expectations. Airlines must also provide records of meeting or exceeding these requirements at regular intervals. All of this information will be requested in the event a safety incident or accident occurs.

DELPHI ANALYTICAL HIERARCHY PROCESS MODEL

There are several methods that can be used to determine SPIs, including the analytical hierarchy process (AHP) and the Delphi method, together known as the DAHP model for quantitatively measuring SPIs. The AHP starts with identifying the top-level objective, which for the commercial airline example may be *minimizing the occurrences and effects of safety related incidences or accidents*. The hierarchy is then established, identifying child objectives for safety risk, cost, benefit, regulatory, and customer expectations. Shown below in figure 6.1 is the typical matrix method for displaying the AHP elements (E_n) using a nine-scale approach, providing a quantitative approach to determining the priority of SPIs.

Using the Delphi method provides a means of determining and prioritizing SPIs using qualitative methodology. Industry experts get together to determine appropriate objectives and indicators based on their extensive

$$\mathbf{A} = \begin{array}{c} \\ E_1 \\ E_2 \\ \vdots \\ E_n \end{array} \begin{array}{c} \begin{array}{cccc} E_1 & E_2 & \dots & E_n \end{array} \\ \begin{bmatrix} a(1,1) & a(1,2) & \dots & a(1,n) \\ a(2,1) & a(2,2) & \dots & a(2,n) \\ \vdots & \vdots & & \vdots \\ a(n,1) & a(n,2) & \dots & a(n,n) \end{bmatrix} \end{array}$$

Figure 6.1 AHP Matrix Model.

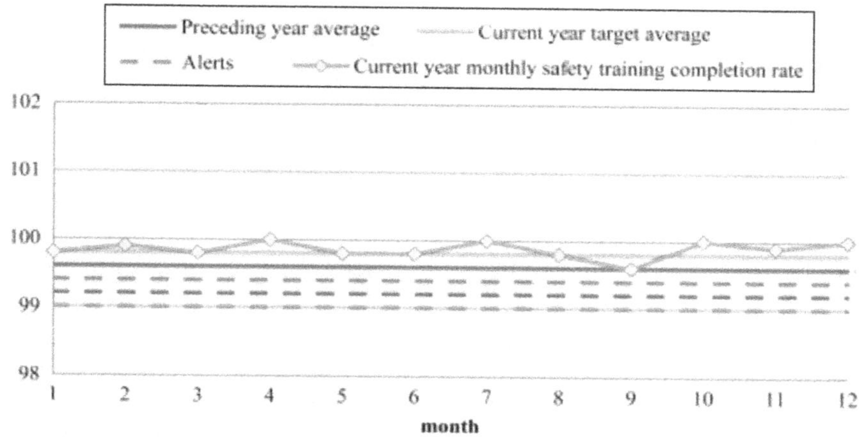

Figure 6.2 DAHP Model Output for Safety Training Completion Rate SPI.

experience. Their opinions are established by use of a questionnaire. The experts are qualified based on their knowledge of safety performance and can come from various parts of the aviation industry. Together, they weigh the value of each of the SPIs to determine the respective SPI position in the AHP matrix. The DAHP model combines the qualitative expertise of industry professionals with the consistency and reliability of a known mathematical model for weighing organizational objectives. An example output of using the DAHP model to evaluate an SPI of safety training completions, a regulatory requirement, is shown below in figure 6.2. This methodology could be applied many different applications within the SMS.

Airlines and other organizations may use other similar analytical forms, or may use committees and company analysts, to determine their specific safety-related goals for a given timeframe. The overall purpose remains the same, to define a set of measurable safety objectives, establish their priority (high consequence versus low consequence), alert levels, and acceptable target levels.

ORGANIZATIONAL PERFORMANCE

Once the SPIs, as well as their target and alert levels, have been defined and accepted by an organization, the next step is to monitor the organization's performance against the SPIs on a regular basis (ICAO, 2013). At some greater time period, such as fiscal quarters, the organization may find it useful to compile an aggregate summary of performance from the previous months to present and evaluate any trends that may have emerged requiring corrective action. So how does an organization go about monitoring its performance within an

SMS? Perhaps it is best to discuss using an example from a small regional airport faced with the challenge of being competitive in the current commercial aviation environment with a high number of large, international airports.

CASE STUDY: PERFORMANCE MANAGEMENT AT SMALL REGIONAL AIRPORT TO IMPROVE SAFETY

The goal was to improve organizational performance management while increasing the safety of operations using a systematic framework and new information technology (IT) system in accordance with the ICAO SMS Manual (Leva, Del Sordo, & Mattei, 2014). The IT system was the foundation of their improvement, which provided a means for monitoring, tracking, and analyzing day to day performance of key airport activities. It supported the development of a formal SMS for the small regional airport, which included provisions for performance monitoring, reporting, support, and documentation of risks. Steps for improved organizational performance management included structured interviews and observations with stakeholders, analyzing documentation, development of focus groups searching for internal improvements and brainstorming new ideas (Leva et al., 2013).

After the review was completed, it was determined the airport was missing two key items: monitoring of crew staffing to ensure enough people with the required skills were retained and proper assessment and feedback of individual employee performance (Leva et al., 2013). Recommended changes included establishing a set of flow down meetings for information transfer between levels of management to the workers, using databases to monitor daily performance data and manage airport equipment and infrastructure, clarifying organizational structures with clear supervisory roles, providing a better communication system, and improving team work with individual responsibility, accountability, and authority (RAA) documented in writing (Leva et al., 2013).

To improve performance management, in addition to addressing the employee recommendations, the airport defined performance indicators at the macro- and micro-task levels, defining the importance of a microlevel task with respect to the macrolevel tasks; this corresponds back to the AHP hierarchy and DAHP model, without formally being called such. Score cards were created to track measurement of the key performance indicators using new software, which also included an organizational chart of employees showing their reporting structure, RAA, and workload, both current and forecasted (Leva et al., 2013). To tie everything back to the SMS, the software also included a hazard analysis tool for each discipline. A visual depiction of the software is shown in figure 6.3.

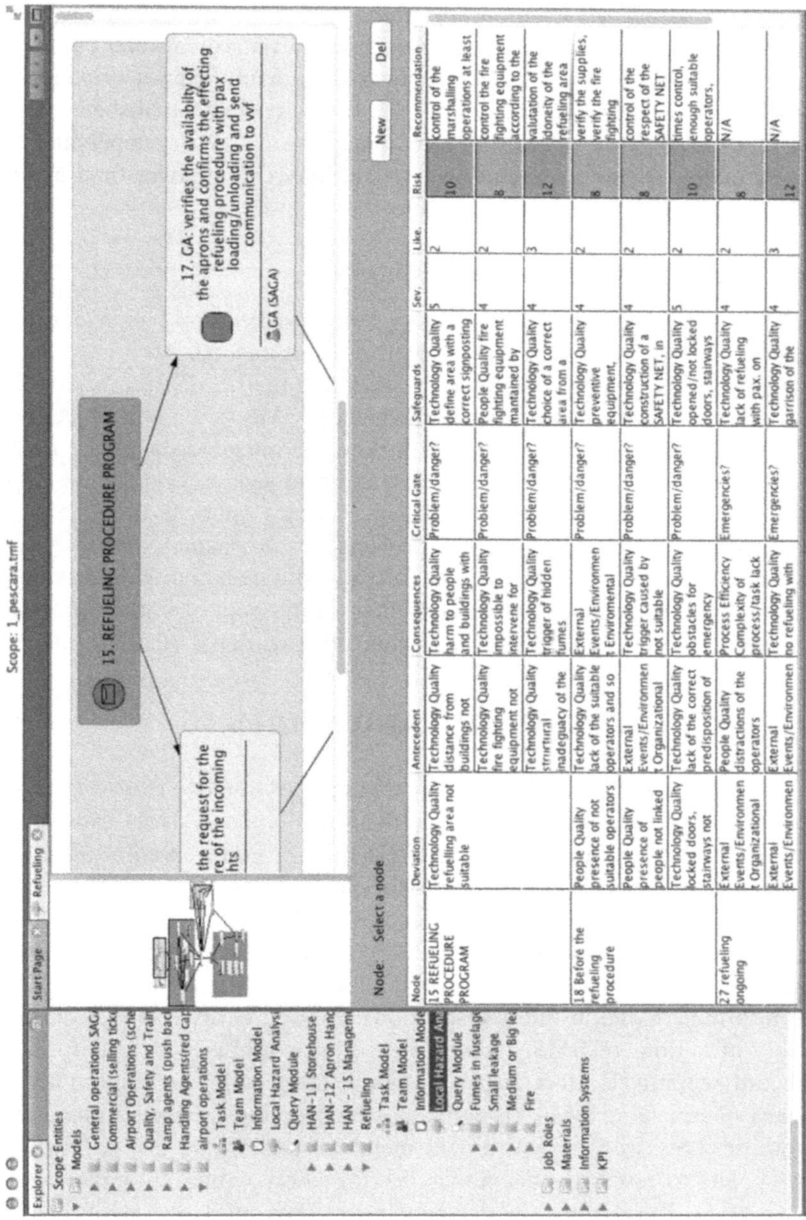

Figure 6.3 New Software Development to Improve Performance Management at Small Regional Airport.

Overall, the key takeaway from the case study of the small regional airport is how organizational performance management is a function of three key items: understanding what the organizational goals are, understanding how the organizational goals specifically apply to each individual, and having a standard set of processes and tools in place to document their performance to plan at predefined intervals. The complexity of the system established within an SMS for the purposes of performance management can vary depending on the needs of the organization and project and may change over time as the organization, project, and industry mature.

Performance Data Collection and Metrics

Part of developing an SMS is establishing a set of metrics by which an organization's safety performance can be measured and tracked. An organization first establishes the goals and objectives pertaining to its SMS and then determines the best way to monitor its performance. SMS goals and objectives vary between organizations, but overall, they are intended to reduce risks, reduce the number of incidents, accidents, and near misses, and decrease the impacts associated with risks and incidences. Metrics can be used to measure the performance of individuals, teams, projects, or the organization. They can be used for personal accountability and are also often shared with organizational management, leaders, and stakeholders. At the root of it, their purpose is to answer the question, "How safe is your organization?" (Eiff, 2000).

INSPECTIONS AND AUDITS

Organizations implementing an SMS are taking steps to move from a reactive to a proactive safety culture where information is collected, root causes are determined, behaviors are altered, and real time progress monitoring takes place (Eiff, 2000). It is important to note for objectives to be measurable, they must be observable through physical action or documented reports, easily identifiable, and not ambiguous (Eiff, 2000). Safety-related data is collected through various means, including inspections and auditing, which can be formal or informal. Aviation safety inspections are used to discretely assess the reliability of various of SMS, such as aircraft airworthiness and safety incident record keeping (Britton, 2016). Aviation safety audits are a more holistic means of assessing SMS compliance by observing how well processes are working, how well organizational members understand and follow the processes, and to assess the accountability for safety within an organization (Britton, 2016). The results of both safety inspections and safety audits can be documented using audit log narratives and inspection checklists. Examples of safety inspection checklists, are shown below in figure 6.4.

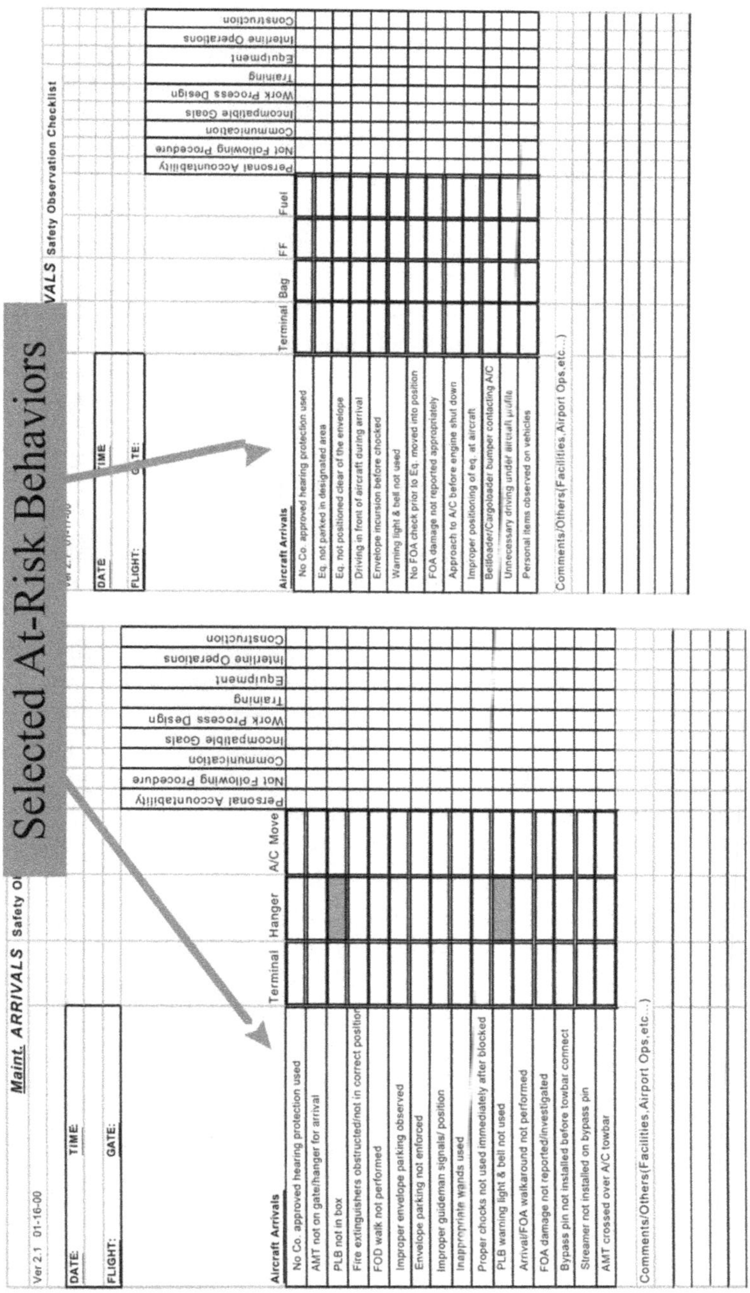

Figure 6.4 Example of Safety Inspection Checklist.

Occurrences of specific incidences can be plotted over time and by a given entity, such as an individual or team. Using a tool such as Microsoft Excel to populate the data into the safety observation checklists and compiling multiple reports at a time can result in a performance metric similar to the one depicted in figure 6.5.

After completing safety inspections and audits, monitoring them over time and with respect to different individuals or teams, patterns may be detected that indicate an existing or emerging safety-related problem. Since the organization took the step to keep detailed records of inspections and audits, and tied them back to specific entities and timeframes, a root cause analysis can be performed resulting in a corrective action plan back to the entity where the problem first emerged (Britton, 2016). Inspections and audits are both required in root-cause analysis. An inspection provides real-time, detailed data on a given occurrence, and the audit brings visibility to where potential breakdowns exist in the process.

PERFORMANCE REVIEWS

Unfortunately, safety inspections and audits cannot effectively account for certain humanistic elements within an organization that may negatively impact safety performance, such as poor relationships between management and employees, the cultural mindset and behaviors regarding safety, and determining whether inspections are thorough enough (Britton, 2016). In these cases, it is imperative for individual performance reviews to be completed between peers, managers, and their direct reports. In these performance reviews, the individuals can assess the individual's or team's performance and work together to qualitatively assess any humanistic elements negatively impacting the organization's safety performance; for example, if an employee feels pressure from his or her manager to perform a task more quickly, thus not following safety procedures, the disconnect can be discussed and resolved during a performance review. It is important the manager creates an open culture where the employees feel comfortable providing feedback to and asking questions of their manager if they are feeling pressure to compromise safety in any given situation. Unless malice is involved requiring disciplinary action, feedback from a performance review should not result in fear or punishment; rather, the performance review should be seen as an opportunity to learn and improve (Stolzer & Goglia, 2015).

SIX SIGMA APPLICATION TO SMS

Mature organizations may take implementation of an SMS a step further than FAA and ICAO standards by also integrating lean six sigma practices. Six

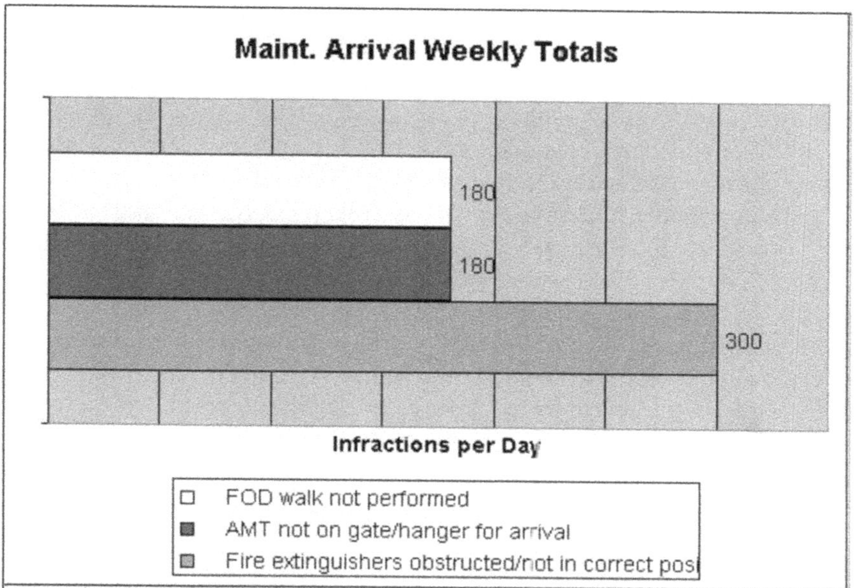

Fire extinguishers obstructed/not in correct p	
Root Cause Percentages	
Personal Accountability	0%
Not Following Procedure	0%
Communication	0%
Incompatible Goals	0%
Work Process Design	0%
Training	0%
Equipment	0%
Interline Operations	0%
Construction	100%

AMT not on gate/hanger for arrival	
Root Cause Percentages	
Personal Accountability	0%
Not Following Procedure	0%
Communication	0%
Incompatible Goals	0%
Work Process Design	0%
Training	100%
Equipment	0%
Interline Operations	0%
Construction	0%

FOD walk not performed	
Root Cause Percentages	
Personal Accountability	0%
Not Following Procedure	0%
Communication	100%
Incompatible Goals	0%
Work Process Design	0%
Training	0%
Equipment	0%
Interline Operations	0%
Construction	0%

Figure 6.5 Example Safety Performance Metric Output.

sigma is a data-driven, quality improvement process widely used in the aviation industry. To be *lean* means to improve speed and efficiency of processes; implementing lean six sigma combines these two processes to improve quality and efficiency simultaneously (Furst, 2010). Using the data-driven approach provided by six sigma, organizations can more easily manage their safety goals and objectives. Six sigma metrics are created in alignment with its three key elements: customer focus, internal processes, and organizational staff (Furst, 2010). Each element has two corresponding pillars that add up to speed, quality, flow, variation control, teamwork, and learning (Furst, 2010). Based on these elements, metrics can be created within an SMS to measure staff training regarding safety policies and procedures, the amount of change in process flow times, and the quantity of quality defects as a result of implementing safety-oriented policies and procedures. Six sigma provides an established framework for integrating safety-oriented policies and procedures into metrics for communicating relevant information to leadership and stakeholders.

Performance Improvement

The audits and inspections have been completed, the root causes have been identified, the hazards and risks have been classified, and organizational performance is being tracked—so how does performance actually improve? The key is to use the SPIs to drive real, tangible, and sustainable change (Ulfbengren & Corrigan, 2015). Additionally, the SPI-driven change process should be linked to the risk management process established as part of the SMS (Ulfbengren & Corrigan, 2015). As previously mentioned, part of defining SPIs is also defining the target and alert thresholds. In the event the results of an inspection or audit and their performance over time indicate the exceeding of an alert threshold for a given SPI, a root cause investigation is completed, which subsequently results in a corrective action plan being created (Ulfbengren & Corrigan, 2015). The organization must create a process by which corrective action plans are documented and followed through. The plan should have milestones, deadlines, and action owners who are held accountable in accordance with a standard change progress used by the organization. In the case of human-related corrective action plans, it is important to maintain a culture of accountability instead of blame. Corrective action plans for personnel accountability should be formal, specific, and comprehensive (Stolzer & Goglia, 2015).

Corrective action plans come at a cost and, as such, organizational leadership must have a strategic means of making decisions regarding corrective action plan development and implementation (Ulfbengren & Corrigan, 2015). In many cases, there could easily be more than one solution to the root cause

Figure 6.6 Safety Risk Management Continuous Improvement Cycle.

problem, each varying in cost and length of time to implement. The leadership team should use established thresholds for cost and flow targets, against which corrective action plan proposals can be evaluated to determine a path forward (Ulfbengren & Corrigan, 2015). Once a corrective action plan is put into place, the audits, inspection, performance reviews, and metrics reporting proceed as a means of evaluating operations and processes for continuous improvements. Shown below in figure 6.6 is a chart depicting the safety risk management cycle. The SPIs have measurable targets which are tracked in the form of reports and data. Whenever SPI alert thresholds are crossed, risk analysis and assessment are completed, which drives a corrective action plan equating to sustainable change.

CONCLUSION

Establishing a safety management within an organization, especially a large organization such as a commercial air carrier or large aircraft manufacturer, is no small task. It requires months of planning and development with participation from organizational members, external stakeholders, and industry experts. After all of the building blocks are in place, successful implementation of an SMS requires a means of continuous improvement comprising safety performance indicator measurement, performance management of individuals and teams, and a lean, effective change management process. There are many tools available to organizations implementing performance management as part of their SMS efforts, including the Delphi Analytical Hierarchy Process Model, Lean Six Sigma, risk cards, metrics, checklists, audits, and inspections. Having a performance feedback loop for personnel, procedures, and tools, with the ability to measure and track their improvements is the key to success when implementing a safety management system.

REFERENCES

Britton, T. (2016). Real difference between and aviation safety audit vs. inspection. *Aviation Safety Blog*. Retrieved from http://aviationsafetyblog.asms-pro.com/blog/aviation-safety-audit-inspection-difference.

Chen, W., & Li, J. (2016). Safety performance monitoring and measurement of civil aviation unit. *Journal of Air Transport Management, 57,* 228–233.

Eiff, G. (2000). At risk safety metric. *Federal Aviation Administration.* Retrieved from https://www.faa.gov/about/initiatives/maintenance_hf/library/documents/media/human_factors_maintenance/at-risk_safety_metric.pdf.

Furst, P. G. (2010). Lean six sigma innovative safety performance management. *Injury Prevention, 16*(1), 92–93.

International Civil Aviation Organization. (2013). *Safety Management Manual.* Retrieved from https://www.icao.int/safety/SafetyManagement/Documents/Doc.9859.3rd%20Edition.alltext.en.pdf.

Leva, M. C., Del Sordo, D., & Mattei, F. (2015). Day-to-day performance management in a small regional airport and management of change for safer operations. *Cognition, Technology & Work, 17*(2), 237–248.

Stolzer, A., & Goglia, J. (2015). *Safety Management Systems in Aviation.* Burlington, VT: Ashgate Publishing Company.

Ulfbengren, P., & Corrigan, S. (2015). Development and implementation of a safety management system in a lean airline. *Cognition, Technology & Work, 17*(2), 219–236.

Chapter 7

Managing Human Factors

Susan K. Archer and Stacie L. Fain

SMS is only as effective as its component parts, and these parts are effective only when members of the organization apply SMS processes to the work they do. Understanding the interrelation between human factors and safety management is paramount in implementing a successful SMS. The study of human factors examines the human interaction in the environment with processes, products, equipment, and facilities (Sanders & McCormick, 1993). Sanders and McCormick adapted a definition of human factors that was first presented by Chapanis, "Human factors discovers and applies information about human behavior, abilities, limitations, and other characteristics to the design of tools, machines, systems, tasks, jobs, and environments for productive, safe, comfortable, and effective human use" (Sanders & McCormick, 1993, p. 5). Although the terms *human factors* and *ergonomics* are sometimes used interchangeably, it is valuable to recognize that ergonomics is concerned with the physical aspects of the working environment, such as specific tool placement; whereas, human factors is generally concerned with the psychological factors. Those who study human factors have two main objectives: (1) to enhance the effectiveness and efficiency of the work that is being done and (2) to enhance human values, such as reducing stress and fatigue, increasing safety, and improving job satisfaction (Sanders & McCormick, 1993). These objectives are inherently tied to effective use of SMS principles.

This chapter begins with a brief history of human factors' evolution. The majority of the chapter examines four critical human factors areas as they pertain to aviation and industry workplaces and SMS: (1) ergonomics-man/machine interface and human efficiency, (2) culture and psychosocial issues, (3) physical factors, fatigue, and human frailty, and (4) musculoskeletal and cumulative trauma disorders. The chapter closes with current guidance status and offers potential solutions for common problems in labor intensive operations and the general work environment.

EARLY DEVELOPMENT

Early studies linking humans and machines have generally been viewed from individual discipline perspectives: the study of the human body or the study of how a machine works. One of the first historical figures recognized for integrating the study of human-machine interface was Leonardo da Vinci (Baeck, 2008). Baeck (2008) noted that da Vinci recorded functional measurements of the human body in motion and then used those measurements for designing machines. In da Vinci's time, machines were generally powered by human movement, so his study of human dynamics was critical to improving the overall design of the machine and its operation. Da Vinci's work also indicated an understanding of the effect of human factors on productivity. According to Baeck (2008), da Vinci likely used time and motion studies to examine how long it would take to haul soil for a deep canal construction project. The previously used method of having workers each use a shovel to dig at different points could have led to dirt slipping back down the slope and into the canal, as well as increased worker injuries from proximity and needing to balance on a slope while working. Da Vinci proposed having fourteen workers pass buckets of dirt up the slope, much like the bucket brigades that were used to fight fires in the U.S. centuries ago.

Many of da Vinci's designs for manned flight also included indications that he incorporated the study of human factors in his designs. Baeck (2008) stated that "there is enough evidence in his . . . designs that his interest in safety and comfort lies well beyond his own" (p. 8). This evidence included recommendations for redundant wing systems to mitigate the risk associated with a wing failure, integration of a shock absorbing bag system to prevent pilot injury in the event of a mishap, and consideration of a pilot's comfort in recommending the use of chamois skin for parts of the machine on which the individual would have to rest (Baeck, 2008).

The physical impact of the human-machine interface has been more formally studied and described for over 300 years. Bernardini Ramazzini, the father of occupational medicine, wrote about physical outcomes related to work exposure with a focus on musculoskeletal disorders (Gainer, 2008). Ramazzini recognized that worker injuries were due to unnatural postures, body movements, and prolonged work in static positions. Fixed working postures and repetitive motions were directly linked to repetitive strain injuries (Gainer, 2008).

IMPACT OF THE INDUSTRIAL REVOLUTION

The advent of the Industrial Revolution brought further light to these types of injuries as workers left farms and began working in factories where they executed repetitive motions in awkward positions for many hours. Workplaces were sometimes extremely dangerous due to existing environmental hazards and less than optimal working conditions. Studies of the workplace generally focused on improvements to increase production and income rather than advancing the human or safety environment. Frederick Taylor's time and motion studies in the early 1900s introduced standardized work practices through scientific examination of work processes. Taylor broke manufacturing processes into component parts and then investigated each part to determine the most efficient and effective methods for accomplishing the tasks. These methods essentially turned the factory line-worker into part of the assembly line, with little concern for his or her physical or emotional well-being. Taylor believed that management should do the company's thinking while the workers should simply do what management decided they needed to do. Little credence was given to *what might go wrong* in a manufacturing process because it was broken into many component steps to minimize the number of actual movements or tasks the worker needed to learn and then do.

Frank and Lillian Gilbreth studied fatigue, design of work stations, and design of equipment for the handicapped in the early 1900s (Caramela, 2018). Their work in time and motion studies (figures 7.1) was considered to be a forerunner to what we now refer to as human factors research and the process they developed for handing tools from nurses to doctors during surgery is still in use today. At the end of World War II, the U.S. Army Air Corps (precursor to the U.S. Air Force) and the U.S. Navy developed engineering psychology laboratories. Dunlap and Associates also began some research as did some companies in Great Britain. The human factors profession was created around 1945 (Sanders & McCormick, 1993).

POST-INDUSTRIAL REVOLUTION DEVELOPMENT

Human factors developed as a formal discipline with the advancement of technology during the twentieth century. Until about 1960, the human factors discipline was primarily researched by the military and the industrial complex. In the early 1960s, NASA began to explore human factors as part of the race for space. The human factors profession saw rapid growth between 1960 and 1980. The computer era brought the study of human factors to the forefront as everyone wanted to develop ergonomically correct computer workstations (Sanders & McCormick, 1993). There were also three

Figure 7.1 Frank Gilbreth Filming a Motion Study, c. 1916.

technology-related disasters in the 1980s that highlighted the lack of attention to human factors as contributing to the disasters: (1) the nuclear incident at Three Mile Island in Pennsylvania, (2) the leak at Union Carbide in Bhopal, India, and (3) the explosion at Chernobyl in the Soviet Union (Sanders & McCormick, 1993). The study of human factors in the United States continued through the late 1900s with a 1988 mandate from Congress for the FAA to expend more resources on human factors research to improve aviation safety. Additionally, the OSHA examined ergonomics regulations for general industry during the 1990s (Sanders & McCormick, 1993). In the years following, several organizations began providing guidance and enforcement for human factors safety-related issues.

ERGONOMICS-MAN/MACHINE INTERFACE AND HUMAN EFFICIENCY

Ergonomics, the European cousin of human factors, developed as a discipline after World War II. Classic definitions included the scientific study of human work, and the application of scientific information concerning human beings to the design of objects, systems, and environments (Gainer, 2008). A more recent description of ergonomics defined it as the theoretical and fundamental understanding of human behavior and performance in purposeful interacting sociotechnical systems, and the application of that understanding to design of interactions in the context of real settings (Gainer, 2008). This definition dovetails the ICAO definition of SMS: an organized approach to managing safety, including the necessary organizational structures, accountabilities, policies, and procedures (Stolzer, Halford, & Goglia, 2008). Managing safety compels one to understand the behaviors of the humans who comprise the organizational structure.

As the twentieth century progressed, the view of an industrial engineer as a manufacturing scientist evolved to place increasing emphasis on studying the human component. In the late 1980s, Alan Hedge, a Cornell ergonomist, applied human factor analysis to the development of forklift trucks based on the operator's needs. Hedge introduced a strategy to develop machine designs that would incorporate three principles: (1) maximize comfort, (2) minimize the risk of accidents, and (3) maximize productivity (Lang, 2004). This design program included a series of steps similar to the steps in the process of developing and implementing SMS. Hedge began with a detailed photographic analysis of existing forklift features that might cause ergonomic problems for workers—much like a team of subject matter experts (SMEs) might examine a new design to identify possible hazards. Hedge asked the focus groups of truck operators to evaluate prototype new vehicles from manufacturers and examined the economic impact of changes to the machinery, similar to a risk assessment. Hedge then developed an ergonomic audit worksheet and used data from a variety of additional sources to create a comprehensive information base for designing forklifts that would "accommodate a wide range of operators, in terms of body size and strength and ease-of-use . . . [paying] careful attention to the ergonomi[c] design of controls, displays, and compartments" (Lang, 2004, p. 22). This is similar to the way an SMS team must consider the human-machine interface to effectively *engineer out* potential hazards.

Another formal study of human-machine interface and the effect on efficiency was the work completed by Chaffin, Faraway, Zhang, and Woolley (2000). The researchers looked at how diversity of individual characteristics might impact reach motion in vehicles and office workstations. The intent of the study was to identify differentiating characteristics based on size, age, and gender that could be used in the creation of simulation software packages for designers and manufacturers. Chaffin (2007) proposed that valid human posture and motion prediction models could be combined with psychophysical and biomechanical models to forecast dynamic human activity and limitations specific to population subgroups. This research can provide valuable data to manufacturing firms for their product development. If a company intends to expand into a particular international market, ensuring that the equipment purchased is ergonomically efficient for the employees in the region of expansion would follow the SMS guideline of *designing out* potential hazards.

EXAMINING THE AVIATION COCKPIT—HUMAN FACTORS AND SAFETY MANAGEMENT

The primary workplace for aircrew members is the cockpit. Morant's (1955) seminal work in anthropometric research found the following six body measurements had the strongest correlation with cockpit design: (1) stature, (2) sitting height, (3) arm length, (4) thigh length, (5) leg length, and (6) seat breadth. Seat location with respect to the instrument panel and flight controls, as well as the limits for mobility control of the seat itself, can create a hazard for aircrew members who are either significantly larger or significantly smaller than the average pilot. Stewart (1955) performed a similar study of cockpit ergonomics; the findings showed that relative size of individual crew members could have a dramatic impact on their ability to fly the airplane safely. Stewart (1955) recommended that aircraft manufacturers use standardized cockpit dimensions for: (1) clearance above the head, (2) degree of vision over the nose, (3) position of the sighting line, and (4) position of the instrument panel. Later studies of the man–machine interface in aviation produced similar results (see Bittner, 1976; National Aeronautics and Space Administration, 1978; Croney, 1981; Schafer & Bates, 1988; Schrimsher & Burke, 1989). In more recent years, multiple aircraft manufacturers developing a variety of aircraft have created a related technology hazard. Newer aircraft do not have the same instrument configuration as their predecessors. Pilots who fly more than one type of aircraft in order to perform their work are forced to adjust to different configurations. The hazard is the potential for time delays in responding to critical alarms or warnings because a switch or knob is in a different location than where the pilot expects it to be. Perhaps the pilot uses the wrong switch or knob to perform a particular task. The repercussions of such a mistake could be lethal. While a pilot is trying to locate a switch or knob that is in a different location in a different model aircraft, his or her situational awareness is diminished. When situational awareness is diminished, an incident or accident is more likely to occur.

CULTURE AND PSYCHOSOCIAL ISSUES

The discussion of culture as it applies to human factors and SMS is twofold: one should examine the organizational culture to determine the significance level of safety management to the organization, and investigate how the cultural backgrounds of employees affect the psychosocial aspect of their participation in and contributions to SMS.

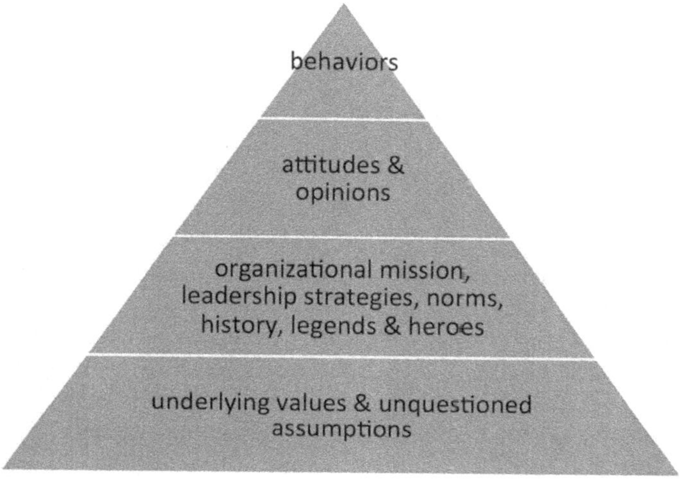

Figure 7.2 The Safety Culture Pyramid.

Developing a safety culture, a critical component of a successful SMS, rests somewhat on the existing organizational culture of the particular company. Patankar and Sabin (2010) described safety culture as a pyramid (figure 7.2). The base of this pyramid was comprised of underlying values and unquestioned assumptions: What is the true basis for the organization's commitment to safety? The next layer of culture included the organizational mission, leadership strategies, norms, history, legends, and heroes: How does the organization narrate its commitment to safety? Based on these two layers, the next layer included attitudes and opinions: What do management and employees believe about the organization's commitment to safety? The top of the pyramid was designated behaviors: How do management and employee actions manifest the first three layers of the pyramid? If an organization intends to implement SMS, it is imperative that leadership collaborate with workers to examine all layers of the Patankar/Sabin pyramid, to understand where the organization *is currently* before trying to achieve where it *wants to be.* This important step must take place prior to any measures to implement SMS.

Why is a safety culture important? From safety professionals to business leaders to students, we are all aware of the apparent dichotomy of economics and safety. Investing in safety does not yield an easily calculated increase to a company's financial bottom line. In the aviation industry, increasing seats and routes, modernizing the fleet by adding or converting to more fuel-efficient aircraft, improving customer relations through enhanced training are costs that commonly result in increases in ticket sales and greater profitability. Investing in employee training programs to implement an SMS results in a more indirect profit: the reduction of costs due to mishaps, accidents, and incidents.

How do we build a safety culture? It starts with the highest level of commitment to safety by top management. Once management demonstrates a commitment to safety, this same level of awareness will trickle down to the lowest level employee. Without the commitment and support of top management, an effective safety culture cannot exist. Supplemental to the support of top management, the organization attempting to revamp its safety culture would benefit from investing in employee training to emphasize the importance of having a strong safety culture and why it should matter to them to employees. Investing in comprehensive training to infuse a safety culture may be costly, but the eventual results should replicate outcomes similar to those of Deming's quality circles culture embraced by Japanese manufacturing firms after World War II. Considering every employee as an SME within his/her area encourages a free flow of information further fostering the trust and teamwork necessary to keep an organization's safety culture strong and healthy.

In today's global economy, it is becoming more likely that a company will either have international components or will interact with stakeholders from other nations. Even within a domestic environment, a U.S. firm could expect employees from a variety of cultural groups. Fassinger (2008) noted that "the strength of the workforce depends on the full utilization of the talents, abilities, and perspective of diverse workers. Research indicates that diversity can be highly effective in workplace tasks requiring innovation and exploration of new opportunities and ideas" (p. 253). Isaac Munene, a senior system safety engineer working in Bell Helicopter's Reliability and Maintainability division, offered the following about a diverse culture:

> A pluralist social culture is where several cultures are recognized and integrated for their unique cultural identities, beliefs and practices. . . . Textron, Inc., like many conglomerates, has offices or business units all over the world. This means it has to have the ability to not only understand and integrate the cultures of their community but also to ensure that everything they do is consistent with the local laws and regulations. Textron, Inc. is the parent company of Cessna Aircraft, Beechcraft Corporation and Bell Helicopter. Bell has subsidiaries such as Bell Helicopter Canada, Bell Japan and Bell Singapore, while all. . . three companies [Cessna, Beechcraft, and Bell] share a common company in Textron India Private Limited. Due to the differences in culture at each location, the organization makes a great attempt at integrating them all without making one feel less significant . . . at every training we are reminded to be aware of all cultures and to educate ourselves on the intricacies of doing business, whether internal or external. (Munene, personal communication, June 16, 2014)

The global venue of aviation requires this multicultural awareness. Within the SMS arena, appropriate training must be designed around the culture of the employees. In certain cultures, it is considered unacceptable to question elders or authority figures. This cultural norm has been cited in reports about the 1997 Korean Air Flight 801 crash, in which the Boeing 747 crashed into the mountains near the destination airport in Guam. A pillar of Korean culture is the belief that "it is . . . more than rude to question an elder" (Davis, 1997). The airline's seniority system at the time assigned to the senior pilot absolute authority over the copilot. According to Leo Mullin, the then chairman of Delta Air Lines, Korean Air's training program in the late 1990s included this authoritative cockpit culture as a major component (Korean Air, 2001). The lack of communication due to cultural norm-imbued training resulted in an accident with devastating effects. When the native cultural norms do not support open communication or a flat organizational structure, SMS training must include emphasis (and perhaps additional training modules) on how to transcend the cultural expectations to ensure that hazards are identified and the risks assessed by team members. These hazards may not just relate to human-human interaction but may also be associated with human-machine interaction.

PHYSICAL ISSUES AND PHYSIOLOGY

A somewhat unanticipated consequence of increasing access for traditionally under-represented groups is the possibility that workspaces traditionally designed to accommodate the average male's physiology may not accommodate female employees or employees from ethnic groups that are generally larger or smaller in stature than the average American male. A related concern involves strength and endurance levels of employees. Roskam (1989) identified four major elements in cockpit design: (1) crew members must be in a position where they are able to reach all controls comfortably, (2) crew members must be able to see all essential flight instruments without undue effort, (3) crew members must be able to communicate by voice or touch without undue effort, and (4) crew members must have visibility outside the cockpit that meets minimum standards for safety of flight. These elements are integral to effective aircraft control, direction, communication, information access, and crew comfort (Archer, Brewer, Fain, & Smith, 2014). In a study of gender differences and general aviation cockpit design (Archer et al., 2014), three of four SMEs identified significant issues in viewability and accessibility. All three of these SMEs were smaller in stature than the average male pilot. They discussed having to adjust seat and rudder pedals to the equipment limits, sometimes being forced to forfeit visibility inside or outside the cockpit in order to perform specific piloting tasks. Applying Endsley's and Jones's (2012) principles of managing rampant featurism through prioritization and flexibility (p. 142) and grouping information based on situational awareness requirements and goals (p. 144), the team discussed the recommendation that aircraft designers consider location of buttons and switches based on their functions and how often they are used (Archer et al., 2014).

FATIGUE AND ITS IMPACT ON HUMAN INFORMATION PROCESSING

One facet of SMS growing into a complete discipline of its own is fatigue risk management systems. The lack of attention that accompanies, or is a direct result of, fatigue can be fatal in high-risk industries such as aviation or industrial safety. This particular issue can be used to explain how human factors impede human information processing, which can lead to significant safety concerns. Consider the (fictional) story of Airman Jones and the FOD Walk (excerpted from Archer, 2013a).

Before each flying day, technicians perform a check for foreign object debris (FOD). At a military location, this check is usually accomplished by a group of maintenance personnel who stand in a line at one end of the aircraft ramp, equal interval distances from each other, so that the entire aircraft ramp is crossed. These men and women then walk toward the other end of the ramp, scanning the pavement in a 180-degree pan back and forth, with the singular purpose of discovering and removing any debris that could cause damage if swept up by an aircraft engine intake and passed through the aircraft engine. Such debris could include a small stone, a piece of rubber, a shard of glass or metal, or something larger. If a single piece of debris is picked up by an engine air intake, it will reach a significant speed as it is processed through the aircraft engine with the air that is critical to engine function. A single pinging pebble can destroy the fast-moving rotors of the intake area or any other moving part within the compression, combustion, or exhaust chambers of the engine. If such a catastrophic event occurs while the aircraft is still on the ground, there is a danger to anyone nearby, but *collateral damage* would be relatively minimal. If, on the other hand, the major damage does not occur until the aircraft is taking off or already in flight, the potential for loss of life and machine is significantly higher.

For a FOD walk or inspection, the primary sensory perception would be via sight. Each aviation maintainer on a military line must be equally diligent in performing the visual scan of his/her path. On some military installations, both the night shift crew and the day shift crew combine to have more *eyes on the ground.* One potential problem is simply the effect of fatigue. Perhaps during the past ten FOD walks, Airman Jones has not found any FOD. Thus, he may have developed a mental filter that the likelihood of finding any FOD is very small; he does not *expect* to find anything. Today he has an important supervisory check that will determine if he is recommended for a skill-level upgrade and he has been studying all night. He is a bit fatigued from a long night out, but he hurries back to the shop to do one more review of his task procedures before the check. The expectation to not find any FOD coupled with fatigue led Airman Jones to miss a small piece of metal on the tarmac. Or, he may actually see the shiny reflection of the metal but, because he is filtering everything through his mental image that there will not be any FOD, he perceives the visual illusion to be the reflection of the bright morning sun off a piece of quartzite in the aggregate concrete of the ramp. Airman Jones is in a hurry, so he does not reach down and touch to double check. Perhaps he has had previous experiences where he did reach down to check something shiny only to discover that it was the quartzite's reflection. Now his mental filter may lead him to ignore such visual cues.

The errors caused by the mental filter and visual illusions are part of Airman Jones' process step in the human information processing model. His brain applies previous experiences or related memories that it has stored to fill the gaps caused by sensory filters. Airman Jones decides not to stop at the sight of something shiny because he applies memories of similar situations to the present situation and erroneously decides to move forward. This decision could have potentially detrimental effects. The example of Airman Jones highlights the downward spiral that may occur when fatigue is introduced into the situation coupled with thinking errors. SMS indicators should address the hazards of human frailty, such as fatigue and mental filtering, as part of a comprehensive analysis of the work performed.

MUSCULOSKELETAL AND CUMULATIVE TRAUMA DISORDERS.

Many researchers interchange the terms musculoskeletal disorder (MSD) and cumulative trauma disorder (CTD) because of the similarity in symptoms. A MSD is defined as a single injury or combination of injuries to the back, upper, or lower limbs where there is damage or disorder to the joints or tissues (HSE, n.d.). These disorders tend to develop over a period of time in the workplace, due to the workplace itself, the activity or movement associated with a particular task, or the repetitive nature of a particular type of work. Recent research on MSDs reflects dramatic concerns for workplace safety. Houvet and Obert (2013) recommended modification of individual work stations and possible reorganization of an entire work space in order to relieve stresses due to repetitive motion, extreme postures, and "intensive efforts in a problematic psychosocial environment" (p. S104). Some of the phenomena recognized as possibly contributing to the increase in MSDs in European countries since the early 1990s included:

(1) the increase in productivity pressures and new organizational methods (globalization, just-in-time methods, quality control, etc.) and their socio-professional consequences (lack of job security, changes in workstations, work intensification, etc.);
(2) work intensification, increase in assembly line work; and
(3) automation of a part of the production process, which increases the share of repetitive manual work with low-amplitude movements performed in poor conditions (Houvet & Obert, 2013, p. S104).

OSHA lists the hazards that can lead to MSDs as: (1) exerting excessive force through manual work with heavy objects or equipment, (2) performing the same or similar tasks repetitively for extended periods of time, (3) working in awkward postures that place stress on the body or being in the same posture for long periods of time, (4) localized pressure into the body part, (5) cold temperatures combined with any of the previously mentioned hazards, and (6) combined exposure to several hazards (OSHA, n.d.). OSHA recommends periodic ergonomic review of a work environment and work practices and procedures. These reviews would constitute the hazard identification phase of SMS. Organizations have dramatically reduced MSDs by applying what are simply SMS procedures to determine effective interventions. According to OSHA, in many cases the reduction in MSD injuries were accompanied by increases to work efficiency and productivity. Therefore, developing and implementing a SMS would facilitate the reduction of MSD injuries and foster improved work efficiency and increased productivity.

CURRENT GUIDANCE AND REGULATIONS

The U.S. Congress created OSHA with the Occupational Health and Safety Act of 1970. OSHA is part of the Department of Labor (DOL) and was designed to assure the safe and healthful working conditions for men and women by setting and enforcing standards by providing training, outreach, education, and assistance (OSHA, n.d.) The Act covers workers in both the public and private sector. Along with many other safety concerns, OSHA is concerned with ergonomic hazards and will cite for violations under the General Duty Clause, Section 5(a)(1) requiring businesses to keep their "workplace free from recognized serious hazards, including ergonomic hazards" (OSHA, n.d.). In addition to a citation, OSHA also has the option of issuing an ergonomic hazard letter to a business that is in violation. The decision whether to cite hinges on: (1)

whether an ergonomic hazard exists, (2) whether the hazard is recognized, (3) whether the hazard is causing or potentially will cause serious harm to a worker, and (4) whether there is a way to eliminate the hazard (DOL). In addition to enforcement, OSHA also provides information on ways for businesses to implement an effective program that can reduce ergonomic hazards.

In addition, the FAA has a robust division of human factors professionals that provides human factors guidance to the civilian (commercial and general) aviation community. This Human Factors division provides comprehensive training on topics such as (1) human factors models, (2) usability, (3) displays, (4) controls, (5) environment, (6) cognition, (7) training, and (8) team performance (FAA). There is also a comprehensive document providing guidance for implementing a human factors program for aviation maintenance and inspection (FAA, 2008). The division provides strict human factors requirements during acquisition for the FAA (FAA, 1999).

CHALLENGES IN AVIATION

SMS implementation within the FAA has been a challenge as it crosses a number of divisions within the FAA; some divisions are advancing with SMS implementation much faster than others (Government Accountability Office, 2012). For this reason, it is likely that integrating human factors with SMS implementation has been a challenge for the FAA. This is evident in that human factor has only been addressed in some SMS implementation guidance documents issued by the FAA, but the discussion has been limited. There is no reference to human factors, human engineering, or ergonomics in SMS Order 8000.369A or SRM Order 8040.4A. There is also no reference in Advisory Circular 120-92A SMSs for aviation service providers. There is one reference to providing training on human factors in Advisory Circular 150/5200-37 SMSs at airports. The Airport Cooperative Research Program (ACRP), under the Transportation Research Board (TRB), however, provides a thorough discussion of human factors during SMS implementation in Report 1 Safety Management Systems for Airports Vol. 2: Guidebook. The Guidebook discusses humans factors as they relate to hazard identification, SRM, safety promotion, safety culture, training, and accident investigation (ACRP, 2009).

LACK OF HUMAN FACTORS/SAFETY MANAGEMENT SYSTEMS INTEGRATION

There are a number of possible reasons for the lack of comprehensive human factors/SMS integration within the aviation industry. Foremost would be the dramatic differences in how company leadership has addressed development and implementation of SMSs within their organizations. Another reason might be the erroneous impression that human factors is considered a *soft science* (psychology, sociology, etc.), whereas the aviation and aerospace industries have historically been associated with *hard sciences* (physics, chemistry, engineering, etc.). Although most people recognized that the human-machine interface was related to aviation accidents as early as the first flying accident investigation, only in recent years have researchers begun to formally integrate human factors/ergonomics findings and recommendations with SMS processes to shift from a reactive function to a proactive/preventive function.

SOLUTIONS FOR COMMON PROBLEMS IN LABOR-INTENSIVE OPERATIONS AND THE GENERAL WORK ENVIRONMENT

The understanding of human factors engineering, or ergonomic design criteria, is used to integrate humans effectively with their workplace systems to achieve simplicity, efficiency, reliability, and safety in the general workplace as well as during intensive operations (Department of Defense, 1999). Understanding these principles will allow for a safe, enjoyable working environment for all employees. The best way to achieve an increased understanding and promotion for the importance of human factors is through the implementation of SMS.

Prior to SMS implementation, management should evaluate the organization's existing safety culture to determine best practices that are in place and where gaps may exist. This can best be accomplished through a Safety Gap Analysis (ACRP, 2009). It is evident that serious accidents occur more frequently in organizations with a weak safety culture. Changing a safety culture overnight is not possible, and an organization cannot just hire a consultant to change its culture. All organizations have an existing safety culture and it is important for leadership to determine what organizational structures and practices cause the culture to be the way that it is.

After the gap analysis is performed and buy-in for the program is established throughout the organization, the SMS team will develop an Implementation Plan. Once the plan is accepted by leadership, SMS implementation can begin. In the component one phase, a Safety Policy is developed by leadership that reflects the organization's commitment to SMS based upon the safety philosophy of the organization. Supporting safety objectives will also be developed. Both of these documents will be presented to all the organization's employees. Roles and responsibilities as well as organizational processes for the SMS will also be defined during the component one phase (FAA, 2007).

A thorough understanding of the human factors associated with an organization is crucial to having a good safety program. This understanding is particularly important in the component two phase of SMS implementation: SRM. During this phase, the SRM processes are defined. These include hazard identification, classifying potential risks associated with the hazards, and establishing risk mitigation plans for each risk (FAA, 2007). Triggers that lead to hazards and risks will be evaluated during the component two phase. Investigations into any accidents and incidents will also be conducted (ACRP, 2012).

In the component three phase, safety assurance, self-auditing, external auditing, and safety oversight are conducted to ensure continuous improvement of the SMS (FAA, 2007). Safety experts agree that as a best practice there must be a formal management review of the SMS. The authors of the OHSMS recommended using the OHSMS cycle (figure 7.3; ASSE, 2012). W. Edward Deming, recognized worldwide for his quality management approaches, designed the model and it became known as the Deming cycle and later evolved into the plan-do-check-act (PDCA) model. The model has since evolved into several versions by several different organizations and authors. One evolution, the OHSMS cycle, provides a process for management to review the SMS to ensure its effectiveness, as well as ensure that all measurements, audits, investigations, and preventative and corrective actions are considered during the ongoing planning process (ASSE, 2012). If a separate human factors engineering program is implemented, this model can be used to evaluate it as well.

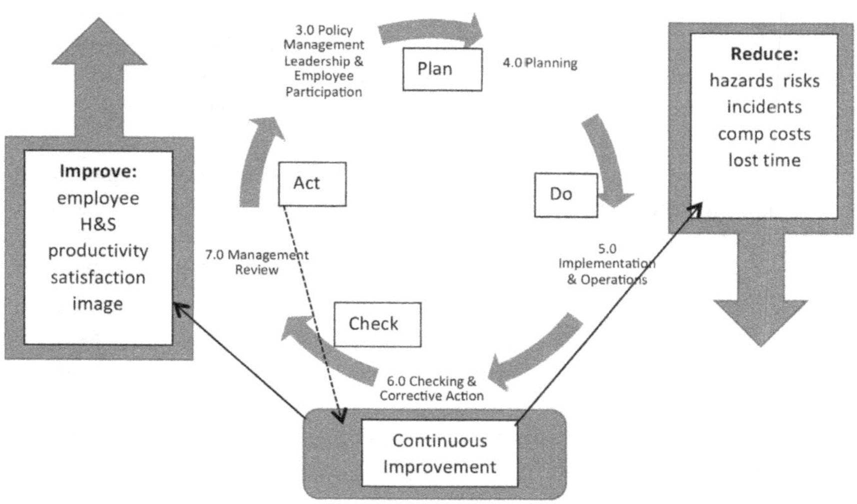

Figure 7.3 Occupational Health and Safety Management Systems (OHSMS) Cycle.

The first step in the OHSMS cycle is Planning. Planning is conducted with the leadership of the organization, management team, and employee participation. Once planning is completed, the resulting proposals are implemented during operation. Operation, when an organization produces its product or service, is the Do step in the process. Evaluation and corrective action are also taken during operations but comprise the Check step in the process. The Act step is where management conducts a review of the process. The model is a circle, indicating that once one reaches the final step (Act) the process begins all over again. This process was designed to facilitate a reduction in safety hazards and continuous improvement of the organization's SMS (ASSE, 2012).

Finally, in the component four phase of SMS implementation, safety promotion, SMS is advertised throughout the organization. This internal marketing and presentation is done through several forums including training and education, safety communication, safety competency, and continuous improvement. Often, safety committees are formed to complete these tasks (FAA, 2007).

Even with a robust SMS in place, accidents and incidents will still occur. It is understood that accidents and incidents usually occur in unusual situations where there is diversion from SOPs. Such diversion is often a result of an increase in operations tempo. These accidents and incidents are usually a result of the operations system and not the individual worker. Manuele (2008) quoted Deming,

> In the American Style of Management, when something goes wrong the response is to look around to blame or punish or look for something to "fix" rather than to look at the system as a whole for improvement. The 85-15 Rule holds that 85 percent of what goes wrong is with the system, and only 15 percent with the individual person or thing (Manuele, 2008, p. 37). The Rule holds that . . . problems in any operation are within the system and the responsibility of management, while only 15 percent lie with the [responsibility of the] worker. (p. 56)

Dr Anita M. Rothblum from the U.S. Coast Guard Research and Development Center stated in her paper, *Human Error and Marine Safety*,

> While human errors are all too often blamed on "inattention" or "mistakes" on the part of the operator, more often than not they are symptomatic of deeper and more complicated system problems. Human errors are typically caused by technologies, environments, and organizations which are incompatible in some way with optimal human performance. These incompatible factors "set up" the human operator to make mistakes. So what is being done to solve this problem? Traditionally, management has tried to either cajole or threaten its personnel into not making errors, as though proper motivation could somehow overcome inborn human limitations. In other words, the human has been expected to adapt

to the system. *This does not work.* Instead, what needs to be done is to *adapt the system to the human.* (pp. 74–75)

Deming's and Rothblum's statements, espousing similar sentiments in consecutive centuries, support the need for understanding human factors engineering and the need for implementation of an SMS that incorporates ergonomic evaluation of SOPs. It also supports the use of the OHSMS cycle to evaluate the processes that have been put in place to ensure robustness and continuous improvement.

R. B. Whittingham, author, chemical engineer, and safety consultant, took the concept a step further and suggested that serious accidents are the result of recurring, but potentially avoidable, human errors. He analyzed several events and concluded that serious accidents were the result of typical system failures that were principally attributed to the organizational, cultural, technical, and management system. D. K. Lorenzo wrote a guide that advises managers on the best practices and "techniques to improve human performance by identifying and eliminating error-likely situations" (Lorenzo, 2001, p. xi). In developing an effective SMS that incorporates comprehensive Human Factors Engineering, it will be necessary to examine SRM from a system perspective vice a worker-risk-management point of view in order to address what will likely cause the majority of accidents and incidents; Deming's 85 percent of what goes wrong in his 85-15 Rule.

In order to address the remaining 15 percent of accidents that are attributed to human error, improvement is needed in the SMS. Dan Peterson captured this idea very well at a 2003 American Society of Safety Engineers (ASSE) Conference:

> For the last 90 years, safety has gone through many frontiers, many fads, and occasionally a paradigm shift. Interest in Behavioral Safety has paled, and some are discovering the importance of management and culture. Although, even in an environment of good management and good culture, people are still being injured due to human error. We need to be able to explain why error happens and what it is. That knowledge will open up the next frontier in safety management. (Manuele, 2008, pp. 67–68)

REFERENCES

Airport Cooperative Research Program (ACRP). (2009). *ACRP Report 1: Safety Management Systems for Airports Vol 2. Guidebook.* Washington, DC: Transportation Research Board.

Airport Cooperative Research Program (ACRP). (2012). *Synthesis 37: Lessons Learned from Airport Safety Management Systems Pilot Studies.* Washington, DC: Transportation Research Board.

American National Standard Institute. (2012). *Synthesis 37: Lessons Learned from Airport Safety Management Systems Pilot Studies.* Washington, DC: Transportation Research Board.

American National Standards Institute (ANSI). (2014). About ANSI. *Overview.* Retrieved from http://www.ansi.org/about_ansi/overview/overview.aspx?menuid=1.

Caramela, S. (2018). The management theory of Frank and Lillian Gilbreth. *Business.com.* Retrieved from https://www.business.com/articles/management-theory-of-frank-and-lillian-gilbreth/.

Department of Defense. (1999). *MIL-STD-1472F Department of Defense Design Criteria Standard – Human Engineering.* Washington, DC: U.S. Department of Defense.

Federal Aviation Administration. (1999). *Human Engineering Program Plan FAA-HF-001.* Washington, DC: Department of Transportation. Retrieved from http://www.hf.faa.gov/docs/did_001.htm.

Federal Aviation Administration. (2007). *Advisory Circular 150/5200-37 Introduction to Safety Management Systems (SMS) for Airport Operators.* Washington, DC: Department of Transportation. Retrieved November 6, 2013, from http://www.faa.gov/documentLibrary/media/advisory_circular/150-5200-37/150_5200_37.pdf.

Federal Aviation Administration. (2008). *Establishing a Human Factors Program.* Retrieved June 13, 2014, from Human Factors Guide for Aviation Maintenance and Inspection: http://www.hf.faa. gov/hfguide/02/02_guidelines.html.

Federal Aviation Administration. (n.d.). Human factors. *Web Training.* Retrieved from http://www.hf.faa. gov/webtraining/intro/intro1.htm.

Health and Safety Executive (HSE). (n.d.). Musculoskeletal disorders. *Manage Risk.* Retrieved from http://www.hse.gov.uk/msd/.

Lorenzo, D. K. (2001). *A Manager's Guide to Reducing Human Error.* Washington, DC: American Petroleum Institute, Inc.

Manuele, F. A. (2008). *Advanced Safety Management: Focusing on Z10 and Serious Injury Prevention.* Hoboken, NJ: Wiley-Interscience.

Occupational Safety and Health Administration (OSHA). (n.d.). Standards and enforcement FAQ. *United States Department of Labor.* Retrieved from https://www.osha.gov/SLTC/ergonomics/faqs.html.

Patankar, M. S., & Sabin, E. J. (2010). The safety culture perspective. In E. Salas & D. Maurino (Eds.), *Human Factors in Aviation* (2nd ed., pp. 95–122). San Diego, CA: Academic Press.

Rostykus, W. (2008). *Ergonomics Program Guidance Document Aligned with ANSI/AIHA Z10-2005.* Fairfax, VA: AIHA.

Sanders, M., & McCormick, E. J. (1993). *Human Factors in Engineering and Design.* New York: McGraw-Hill.

The American Society of Safety Engineers. (2012). *Z-10 American National Standard for Occupational Health and Safety Management Systems (ANSI/AIHA, 2012).* Falls Church: American Industrial Hygiene Association.

United States Government Accountability Office. (2012). Additional FAA efforts could enhance safety risk management. *Aviation Safety.* Retrieved from http://www.gao.gov/assets/650/648110.pdf.

Woodson, W. T. (1992). *Human Factors Design Handbook* (2nd ed.). New York: McGraw-Hill.

Chapter 8

Budgeting for Safety and Risk Assessment

Mark Simpson

How do you evaluate the risk of *not* doing something? That is a fascinating question.

—Zero Dark Thirty, 2012

Safety is everyone's responsibility may be the most overused platitude in the aviation industry. Although the statement has merit, it is still an over-simplified attempt to scope a complex topic into a motivational poster for breakroom bulletin boards across the country. In reality, a successful safety plan benefits employees in more ways than simply reducing on-the-job injuries. With the effective implementation of a safety plan, the business entity, as a whole, will see an increase in productivity and efficiency, as well as a decrease in the unplanned expenses associated with accidents and incidents.

SMS is a requirement for all Title 14 of the Code of Federal Regulations (14 CFR) part 121 certificate holders, per 14 CFR part 5 (2017). The maritime and railway industries have also implemented similar safety management programs. Smaller aviation corporations and other industries, however, are not bound by the same regulations. This does not mean organizations should reject the idea of a robust safety plan, as the benefits are numerous. This chapter will discuss the cost of accidents, budgeting, and the return on investment (ROI) of a successful SMS.

THE COST OF TRAGEDY

Before tackling the cost benefits of enacting and maintaining an effective SMS, one must first address a glaring question: what does an organization stand to lose from a poor safety record? In the aviation industry, the answer is painfully obvious. Trans World Airlines (TWA), under the control of legendary Howard Hughes, was an industry icon at the height of luxury air travel in the 1960s and 1970s. The 1990s, however, brought about financial turmoil for the airline. Many point to the tragedy of TWA Flight 800 in 1996 as the beginning of the end for Trans World Airlines. Ultimately, bankruptcy was declared, and TWA agreed to merge with American Airlines shortly after September 11, 2001. TWA Flight 800 remains one of the deadliest accidents in U.S. history as 230 souls were lost. Following a costly four-year investigation, the National Transportation Safety Board (NTSB) traced the cause of the accident to a fuel tank explosion:

> [An] explosion of the center wing fuel tank (CWT), resulting from ignition of the flammable fuel/air mixture in the tank. The source of ignition energy for the explosion could not be determined with certainty, but, of the sources evaluated by the investigation, the most likely was a short circuit outside of the CWT that allowed excessive voltage to enter it through electrical wiring associated with the fuel quantity indication system. (NTSB, 2000)

The NTSB issued several safety recommendations, mostly aimed at center wing designs and the prevention of ignition sources near the CWT. Unprecedentedly, the high profile disaster altered the public's perception of the airline, and TWA struggled to recover. While Boeing was held responsible for a majority of the $500 million lawsuits following the accident, TWA paid an estimated $27 million to relatives of the victims, not to mention the loss of an $83 million Boeing 747 (Graves, 2002).

Adam Air, an Indonesian low-cost carrier, provides another cautionary tale of the expenses an organization could accumulate in the case of an accident or incident occurring. Adam Air was one of the fastest growing airlines in Indonesia from 2002 to 2008, although rumors of unsafe maintenance practices contributed to its spotty industry reputation. Pilots reported being pressured by management into flying unairworthy aircraft and a number of close calls overshadowed company growth. In 2006, a private equity firm had planned to purchase a 30 percent stake in the airline for an estimated $135 million but withdrew the offer later that year partially due to disagreements over operational and maintenance issues (Greenlees, 2007). Soon after losing a potential investor, an inevitable tragedy occurred on January 1, 2007, when a Boeing 737-400 belonging to Adam Air crashed into the Indian Ocean. The final report concluded the cause of the accident to be pilot error, as the pilots inadvertently disconnected the autopilot while troubleshooting (Greenlees, 2007). When Adam Air had another incident a few months later resulting in a hard landing and aircraft damage, the airline was subsequently banned from operating within the European Union. This was followed by a complete ban from the Indonesian government the following year. Adam Air declared bankruptcy shortly after.

The rise and fall of Adam Air and TWA serve as a warning to other air carriers. While the two companies were at opposite ends of the spectrum in terms of scale, both suffered the same fate from the direct and indirect costs of accidents. The direct costs are obvious: destroyed aircraft, settlements to victims and surviving family members, and search and rescue expenses. Regardless of airline size, repercussions such as these may take years to recover from financially. However, neither airline was able to overcome the indirect cost of a damaged reputation. In transportation industries, the costs of safety violations are extremely high. Following disasters, expenses are frequently compounded by the scale of casualties and fatalities. Public perception becomes one of the most important aspects of a business. As exemplified by Adam Air and TWA, public distrust, from the perceptions of both investors and passengers, resulted in financial burdens that ultimately led to the airline's demise.

While major accidents inevitably generate headlines around the world, incidents could still have a significant financial impact. As of 2012, airlines only make $164 for every $16,400 they spend on a typical domestic flight, that is, a ridiculously low 1 percent profit margin (Lubin, 2012). Predictable expenses like fuel, salaries, ownership costs, and taxes account for approximately 80 percent of expenses. The remaining 20 percent can be attributed to maintenance costs and miscellaneous expenses. With margins being so low, it is easy to see the damage of unplanned workplace incidents. Reducing the number of workers compensation claims and associated medical expenses can be an easy way to increase profit, but unexpected claims and unplanned maintenance costs can be the difference between growth and bankruptcy. Aviation companies employ personnel in numerous supporting roles mirroring a more traditional business model; dispatchers, baggage handlers, customer support specialists, and corporate divisions are all important elements of a successful business, as a successful safety culture reaches all corners of the business.

INDUSTRIAL APPLICATIONS

Other industries are not exempt from lawsuits either. In 2008, a crane collapsed on New York's Upper East Side, killing two men. The crane company worked on some of the largest projects in the city, including rebuilding the World Trade Center. The owner of the company, James Lomma, prioritized profit over safety. Months before the accident, he authorized the use of a faulty bearing for his 240-foot crane made by an obscure overseas company, fully aware of the possible repercussions. In 2015, a jury required Mr Lomma to pay $96 million in punitive and compensatory damages due to his negligence, causing him to declare bankruptcy on his estimated $40 million in assets (Martin, 2015).

While these cases are examples of worst-case scenarios, smaller scale-accidents can also have significant impacts. Failed inspections from regulatory agencies, such as the OSHA, frequently result in penalties. Willful safety violations can incur fines up to more than $126,000 per incident (OSHA, n.d.), and indirect costs must also be calculated. Lost productivity, replacement employee training, and damaged equipment all influence profitability. In early 2010, OSHA began publishing data on individual companies, including total case rate and days away from work. For companies with an impressive safety record, this can be a powerful tool when looking to get a competitive edge or to air a competitor's dirty laundry (Tarrell, n.d.). As previously stated, public opinion can be a powerful influence on the success of a business.

INVESTING IN SAFETY

The cost of implementing state-of-the-art safety tools can be expensive. FOQA is one tool used by major airlines. It allows the company to monitor several data points to improve safety and efficiency. While pilot distrust is sometimes cited as a drawback, the program has been cited as a valuable way to significantly improve flight safety (Civil Aviation Authority, 2002). According to the GAO, the cost of implementing a FOQA program for an airline with 50 aircraft is approximately $760,000 per year (GAO, 1997). While this figure is outdated, it highlights the significant investment for just one SMS tool. Each aircraft must not only have costly equipment installed, but personnel must be hired to interpret the data for the program to be successful. However, after the initial upgrade and implementation costs, the return on investment begins. The GAO also states the same company investing $760,000 per year on FOQA will realize a savings of $1.65 million per year. This is a total savings for the airline of around $892,000. Not only will accident and incident rates be reduced, but data can be used to increase efficiency and minimize unplanned maintenance. While those direct costs alone demonstrate the ROI, an efficient FOQA program will also lower insurance premiums.

The ROI translates to smaller aviation businesses as well. Okwera (2016) found the cost for implementing an entire SMS for a small company would be between $20,000 and $30,000, with yearly expenditures of $15,000–$17,000 (Civil Aviation Safety Authority of New Zealand, 2012). These costs are minimal compared to the direct cost of structural damage to an aircraft, but the

Civil Aviation Safety Authority of New Zealand (CASA) also estimates the indirect costs of legal claims, short term replacement, loss of equipment, and business reputation can accumulate to become twice as high as the direct costs.

When it comes to reducing accidents and injuries while increasing efficiency and profit margins, top executives across numerous industries are clearly getting the message. The American Society of Safety Engineers published a survey across manufacturing, health care, finance, and construction industries to gauge the perceived benefits of investing in safety. The study found that for every $1 spent on direct costs related to an accident, executives estimated another $3–$5 worth of indirect costs. An accident costing the company $15,000 in direct medical and compensation costs puts the actual total business impact somewhere between $45,000 and $75,000. On the other hand, executives stated investing $1 in workplace safety can *save* the company around $4.41 (National Safety Council, 2013). While this was a perception survey, it is extremely important to understand that successful industry leaders recognize the impact of safety on the business and believe making an initial investment in safety will yield quantifiable financial returns.

Another important indirect benefit of an effective SMS deals with the employees. While it might be difficult to quantify initially, employees who feel safe and valued by their employer will not only increase efficiency but are more likely to remain with the company as well. Southwest Airlines is a perfect example of a healthy corporate culture with an employee-centered focus. This culture extends well beyond safety, but many core principles align with basic SMS concepts. Grubbs-West (2005) lists some loyalty lessons practiced at Southwest that demonstrate this connection. Lessons like "keep 'em learning," "hire attitude-train skills," and "immerse everyone in the culture immediately" are directly tied to safety promotion and safety policy. The policy of putting employees first, customers second, and shareholders third (Martin, 2015) has paid off, as the airline consistently ranks amongst the safest airlines in the country (Chandler, 2014). This is largely thanks to the culture implemented by founder, Herb Kelleher, who prioritizes safety as everyone's responsibility. The commitment to employee-focused safety results is demonstrated in the airline safety commitment: All Southwest Airlines employees, from leadership to frontline employees, are responsible for establishing the highest level of safety in our operation and workplace (Southwest Airlines, n.d.). The commitment to employee-focused safety results in low turnover and high job satisfaction. Profit is impacted as well. Over the first 42 years, Southwest carried around 1.7 billion passengers across 23.9 million trips. The airline is virtually accident free, with only a handful of blemishes from aircraft overrunning short runways and no catastrophic incidents (Chandler, 2014). When employees are encouraged to uphold the highest safety standards, the return on investment is realized through reduced maintenance costs and incidents.

ASSESSING RISK AND ESTIMATING BUDGET

In high-risk industries such as aviation, it is not possible to completely eliminate accidents. Thus, identifying hazards and controlling risks can be one of the most important steps to establishing a safety policy that increases profit and decreases unexpected and costly incidents. An accurate assessment of risk can not only aid decision makers in electing to continue certain types of operations, but it can establish a baseline of fixed business costs. In the aviation industry, *risk* is defined as the assessment, expressed in terms of predicted probability and severity, of the consequences of a hazard taking as reference the worst foreseeable situation (ICAO, 2009). This may differ from a traditional business view of evaluating financial risk, but the ICAO definition can be applied to a business risk assessment perspective to uncover potential safety hazards that could impact the business financially. An example could be the loss of an aircraft or the death of an employee.

The financial impact of injuries and other safety concerns is well-documented in the construction industry, as it is characterized by frequent change, suboptimal working conditions with exposure to noise and vibration, and a lack of steady employment (Pinto, Nunes, & Ribiero, 2011). Due to the above average rate of workplace injuries in the industry, risk assessments are a necessary step in the risk mitigation process prior to beginning a job. For industries where analysts are dealing with uncertain and frequently insufficient data, Pinto, Nunes, and Ribiero (2011) recommend fuzzy rule-based safety analysis and highlight the importance of safety *not* beginning in the field. Rather, the process must begin during the procurement phase. From a financial standpoint, a company in any industry can use a risk assessment to accurately gauge the possibility (and cost) of accidents and incidents occurring. Depending on the outcome, an annual safety budget can be established. If this budgetary baseline is not completed, the company is taking a risk. According to the Civil Aviation Authority of New Zealand (2012), "Financial institutions and insurers are risk averse. They assess lending criteria on risk factors which they have built up over many years. Showing them you understand and are managing your risks will give them greater confidence in your organization" (p. 7). Improving the organization's financial profile increases investor confidence and allows the business to accurately establish predictable fixed costs each year.

ESTABLISHING A BUDGET

As companies accept the benefits of investing in safety, creating an accurate budget will be a priority for the business. One important element in developing this financial plan will be to ensure safety goals align with the goals of the entire company. For example, if an aviation company is investing in a new fleet of unfamiliar aircraft, the safety budget might focus on implementing new software tools to ensure pilots are flying safely and efficiently. Although these tools will most likely have a steep initial cost, the ROI will be realized in the following years. If the business is expanding to smaller, regional markets, the safety budget may want to include pilot training for short runway operations, as well as budget for safety professionals to conduct risk assessments for the new airports. The point of these examples is to highlight that the process of establishing a budget can be very different for each business, but aligning the safety plan with company goals will ensure accurate estimates.

One important element of creating a successful SMS is defining specific goals for the company. Metrics can vary from reducing workers compensation claims to the response time to near-miss reports. Regardless, these metrics will be developed by safety professionals within the company. These goals will need to be supported by the safety budget as well, or the business will inevitably fall short of their goals. Each company will have a unique set of safety targets based on past performance. Thus, allocating the same percentage of the yearly budget to safety might not fit for every company. If safety professionals ensure the budget is created with the company direction and safety goals in mind, leadership can decide how aggressively to invest in the SMS.

DATA COLLECTION AND RECORDKEEPING

The importance of data cannot be understated, as it has been called the *new currency* in business circles. With the increasing complexity of business and safety operations, data is often the only way to effectively track processes. According to Brunelli (2014), decision makers are more reluctant to make decisions based off of feelings and hunches; rather, decision makers prefer to use analytic and quantitative tools to base and analyze their decisions. A successful SMS must focus on gathering data and encourage feedback from employees on safety and efficiency of processes. Safety assurance focuses on this information acquisition. In the aviation community, this is accomplished by anonymous reporting tools like ASAP or software solutions like FOQA. Similar solutions exist in other industries as well. The business case for these tools is simple—investing in data acquisition provides a competitive advantage. The reduction in accidents and incidents, accompanied by the corresponding increase in efficiency and productivity, can be the difference between profit and bankruptcy. This holds especially true in industries with extremely narrow profit margins. Tools like occupational risk assessments and analytic hierarchy processes aid businesses in the quality assurance arena as well. Successful businesses will most likely already have a QMS in place to ensure customer satisfaction. The FAA (2017) stresses the importance of ensuring the QMS does not interfere with the SMS; rather, ensure each system is focused on specific goals. SMS should be focused on the safety of all aspects of the business. QMS focuses on customer satisfaction. In reality, if these systems are designed correctly, they are both tools that can be used to ultimately increase profit.

CONCLUSION

Businesses in every industry have a choice when it comes to investing in safety. The associated costs can be viewed as an expense, and leadership can work to minimize this cost in an attempt to increase profitability. The topics covered in this chapter have illustrated how faulty and dangerous this logic can be. Research has shown the ROI for investing in safety is significant, and industry leaders are listening. While accident avoidance is a difficult cost to quantify, risk assessments and other tools can be used to assess the probability of accidents occurring allowing decision makers to accurately budget for safety management.

REFERENCES

Brunelli, M. (2015). *Introduction to the Analytic Hierarchy Process*. New York City, NY: Springer International Publishing.

Chandler, J. G. (2014, May 15). Southwest's amazing safety record. *Airline Ratings.* Retrieved from https://www.airlineratings.com/news/southwests-amazing-safety-record/.

Civil Aviation Safety Authority of New Zealand. (2012). An introduction to safety management system. *Safety Management Systems.* Retrieved from http://www.caa.govt.nz/sms/#1.

Federal Aviation Administration. (2017). Safety management system components. *Safety Management Systems.* Retrieved from https://www.faa.gov/about/initiatives/sms/explained/quality_and_safety_management/.

Government Accounting Office. (1997). Aviation safety: Flight operational quality assurance programs. *Report to Congressional Requesters.* Retrieved from http://www.gao.gov/archive/1998/rc98010.pdf.

Graves, N. (2002). The damage exceeds $500M: Boeing, TWA shell out for kin lawsuits. *New York Post.* Retrieved from http://nypost.com/2002/07/15/the-damage-exceeds-500m-boeing-twa-shell-out-for-kin-lawsuits.

Greenlees, D. (2007, January 30). Indonesian airline had a string of safety issues. *International Herald Tribune.* Retrieved from http://www.nytimes.com/2007/01/30/world/asia/30ihtadam.4403744.html?pagewanted=3&_r=0.

Lubin, A. L. (2012, June 15). Airlines have an insanely small profit margin. *Business Insider.* Retrieved from http://www.businessinsider.com/airlines-have-a-small-profit-margin-2012-6.

Martin, E. (2015, July 29). A major airline says there's something it values more than its customers, and there's a good reason why. *Business Insider.* Retrieved from http://www.businessinsider.com/southwest-airlines-puts-employees-first-2015-7.

National Safety Council. (2013). The business case for investment in safety- A guide for executives. *Journey to Safety Excellence.* Retrieved from https://www.nsc.org/Portals/0/Documents/JSEWorkplaceDocuments/Journey-to-Safety-Excellence-Safety-Business-Case-Executives.pdf.

Occupational Safety and Health Administration (OSHA). (n.d.). Benefits. *Business Cases for Safety and Health.* Retrieved from https://www.osha.gov/dcsp/products/topics/businesscase/benefits.html.

Okwera, P. (2016). A cost-benefit analysis of safety management system implementation in the transportation industry (Order No. 10146878). *ProQuest Dissertations & Theses Global.* Retrieved from http://search.proquest.com.ezproxy.libproxy.db.erau.edu/docview/1829548962?accountid=27203.

Pinto, A., Nunes, I., & Ribiero, R. (2011). Occupational risk assessment in construction industry – Overview and reflection. *Safety Science, 49*(5), 616–624.

Southwest Airlines. (n.d.). *Southwest Airlines Safety Commitment.* Retrieved from http://studylib.net/doc/8739117/southwest-airlines-safety-commitment-s.

Tarrell, A. (n.d.). OSHA incident rate for individual companies now posted online. *HNI.* Retrieved from http://www.hni.com/blog/bid/71350/osha-incident-rate-for-individual-companies-now-posted-online.

Chapter 9

Accident and Incident Investigation

Abraham Adole

> On April 26, 1986, an explosion at the Chernobyl Nuclear Power Plant in Ukraine led to the radioactive contamination of the surrounding countryside and to radioactive fallout throughout Eastern and Western Europe. In a test of the new Soviet policy of glasnost (openness), Soviet authorities acknowledged the disaster, though only after Western countries had traced the radiation source to Ukraine. —Deputy Press Secretary Larry Speakes, 1989

Accidents and incidents serve as an indication of deficiencies within an organization's SMS. The systematic approach of managing safety within any given industry requires identification, assessment, control of hazards, and associated risks. This is the goal that SRM, one of four components of SMS, is intended to accomplish. When an accident or an incident takes place, it means that currently established controls through SRM were insufficient in preventing such an occurrence; hence, this points to an increased need to continuously monitor and maintain the SMS for optimized effectiveness. The ongoing monitoring and systematic evaluation of safety processes and practices demonstrates how the SMS is performing, provides an opportunity to identify and address new hazards/risks, and ensure compliance to safety regulations. These activities fulfill the purpose of safety assurance, another core component of SMS, and require a process in place for data collection and analysis through audits, inspections, employee reporting systems, and accident and incident investigations (FAA, 2015a).

This chapter will focus on accident and incident investigation as an essential part of safety assurance in SMS. It is important to start off with the role of accident and incident investigation in SMS, including some definitions, and set forth best practices and approach an organization should take in conducting an investigation. Then, we will look into the need for integrating human factors, as well as introduce available tools used in investigating. Using the

1986 Chernobyl nuclear disaster as an example, we will demonstrate the lasting impacts that an accident can have, and describe how one would go about investigating such a case. Finally, we will expand on the role of accident and incident investigations in SMS, as it pertains to incorporation of the investigation results for SMS improvements.

THE ROLE OF ACCIDENT AND INCIDENT INVESTIGATION IN SMS

Accident/incident investigation is conducted to answer the following main questions: Why did the event take place? What contributed to the event? How can similar events or consequences be prevented? We already mentioned that an investigation is part of the SA component of SMS and is closely tied to SRM. SMS being characterized as a decision-making process requires a continuous feedback loop between SRM and SA. While SRM is concerned with controlling hazards/risks based on system design, including its operational and environmental aspects, SA is focused on monitoring the performance of a system and is achieved through incident and accident data acquisition and analysis, as opposed to analysis of system description. Results of an investigation, revealing the causal and contributing factors to an event, are used to implement the necessary new controls, reassess risk through SRM processes, and continue further monitoring for effectiveness (FAA, 2015a). This feedback loop of the safety management decision-making process through SRM and SA is best demonstrated by figure 9.1.

It is important to note that SMS does not necessarily guarantee that all hazards/risks will be addressed by an organization. Instead, effective SMS ensures that the organization is operating at acceptable levels of risk and is using SA processes to demonstrate that. Typically, an organization will prioritize risks based on the likelihood of occurrence and the severity of consequences. Accident/incident investigation provides the information needed for organizations to make that decision, as the results are based on actual real-world events, establishing a maximum probability of occurrence and making the severity of consequences more evident (Stolzer & Goglia, 2015). However, one must first find out about an accident/incident in order to investigate. Therefore, having an effective and a formal reporting system in place is what makes SA possible to begin with.

WHEN TO CONDUCT AN INVESTIGATION

All events, big or small, coming through safety reporting systems or through audits and inspections, should be reviewed by the organization, and those

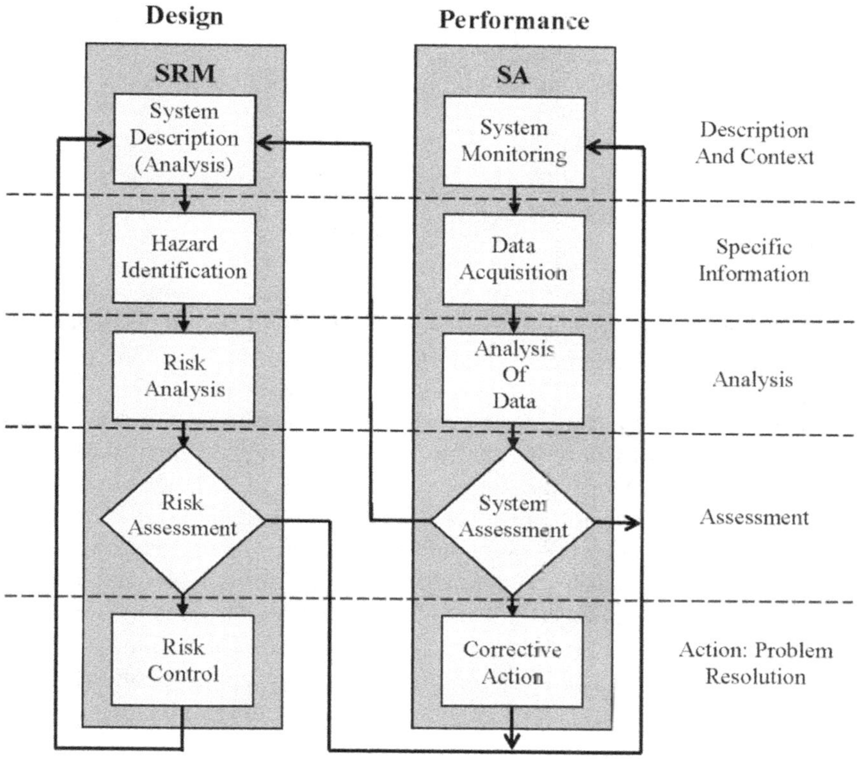

Figure 9.1 Safety Management and Decision-Making Process.

events that have the greatest impact, or potential impact, should be investigated with intent to find the causes and correct them. Reported events could be for different types of hazards (environmental, operational, organizational, procedural, behavioral, etc.), minor mishaps, near-misses, or major incidents/accidents. Even if a reported event does not constitute an actual accident, it should be assessed for the potential to lead to an accident, as minor events and near-misses are often an indication of an accident waiting to happen (Casa.gov, 2014).

How should organizations decide which events to investigate, and to what extent? A common sense approach would suggest that safety events involving some kind of damage to property and/or persons should be investigated. Based on this reasoning, all accidents should be investigated, regardless of the breadth of the damage (Reese, 2012, p. 45). Just as with all accidents, near-miss events should be treated as incidents that require an investigation, as they have high potential for injury/damage. Cases that do not involve an injury or illness but resulted in substantial damage to property should also be investigated (Reese, 2011, p. 24). The extent to which an investigation should be conducted is determined through an initial risk assessment of potential consequences or by the actual consequences of an accident/incident (CASA.gov, 2014).

DEFINITIONS

To better answer the question of when to investigate, it is important to distinguish between various types of events. It is also valuable to understand that one company's approach to safety can significantly differ from another company's approach, solely based on the choice of definitions. Between different industries and regulatory organizations, definitions and associated terminology used to describe safety, and safety-related events can have a significant impact on how safety is being managed. If an organization defines accidents as unplanned events with associated losses and defines safety as lack of accidents, one could rationalize that any event that does not cause a loss, which is the case with many precursor incidents, does not impact safety (Skogdalen & Vinnem, 2012). For the purposes of continuing our discussion on accident/incident investigation, let us examine various definitions in current use by different sectors. Particular attention will be given to distinguishing the different between accidents and incidents for the purposes of this chapter.

Canadian Centre for Occupational Health and Safety (CCOHS)

- Accident—"an unplanned event that interrupts the completion of an activity, and that may (*or may not*) include injury or property damage" (ccohs.ca, 2011).
- Incident—"an unexpected event that *did not cause injury or damage* this time but had the potential. "Near miss" or "dangerous occurrence" are also terms for an event that could have caused harm but did not" (ccohs.ca, 2011).

Washington State Department of Labor and Industries

- Accident—"an unplanned, unwanted, but *controllable* event which disrupts the work process *and* causes injury to people" (OSHA, 2009).
- Incident—"an unplanned and unwanted event which disrupts the work process and has the potential of resulting in injury, harm, or damage to persons or property" (OSHA, 2009).

Health and Safety Executive, United Kingdom (HSE)

- Accident—"any unplanned event that resulted in injury or ill health of people, or damage or loss to property, plant, materials or the environment or *a loss of business opportunity*" (stpaulssteinerschool.org, 2013).

Federal Aviation Administration (FAA) and National Transportation Safety Board (NTSB)

- Aircraft Accident—"an occurrence associated with the operation of an aircraft which takes place between the time any person boards the aircraft with the intention of flight and all such persons have disembarked, and in which any person suffers *death or serious injury*, or in which the aircraft receives *substantial damage*" (e-CFR, 2016).
- Incidents—"an occurrence *other than an accident*, associated with the operation of an aircraft, which affects or could affect the safety of operations" (e-CFR, 2016).

United States Department of Labor

In the past, the term "accident" was often used when referring to an unplanned, unwanted event. To many, "accident" suggests an event that was random, and could not have been prevented. Since nearly all worksite fatalities, injuries, and illnesses are preventable, OSHA suggests using the term "incident" investigation. (OSHA, n.d.)

Based on these varying definitions, it is evident that there is no standardized approach to safety that can be applied across multiple disciplines. As we can see by these examples, in some cases incidents are defined to exclude accidents and are only geared toward events that have the *potential* to cause harm. We can also observe that the extent of injury/damage considered in these definitions of an accident varies greatly, from the lowest level of harm to only serious injuries or substantial damage. There are some common themes across these definitions. First, all accident definitions explicitly state some level of injury or damage. Second, most accident definitions imply some level of disruption of operations/work activities. Finally, definitions of an incident tend to include events that have the potential of causing injury/damage; whereas, to classify an event as an accident, there needs to be some type of loss.

To add to this debate of accidents versus incidents, let us consider one more alternative definition. According to Charles D. Reese, "Accidents are usually defined as unexpected, unplanned, and uncontrollable events or mishaps" that cause some kind of injury or damage, while incidents can be characterized by all of the above, plus include events with "adverse production

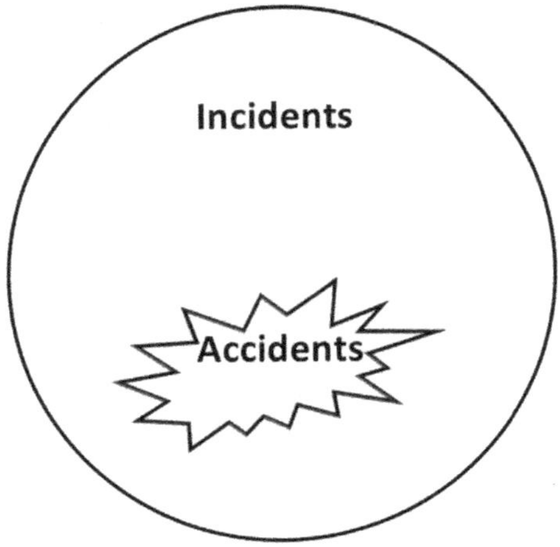

Figure 9.2 Accident as Specific Types of Incidents.

effects" (Reese, 2011, p. 3). Reese proposes to treat accidents as a specific type of incident. This is best demonstrated in figure 9.2.

Going forward, we will adapt Reese's definition of accidents and incidents and use the term interchangeably within the context of our discussions. Thus, we will continue with the position that all accidents are incidents, but not all incidents can be classified as accidents. This model will enable us to approach safety with the philosophy that causal factors to accidents can be controlled, highlighting the value of conducting an investigation.

COST OF ACCIDENTS AND INCIDENTS

Safety Assurance through accident/incident investigation enables organizations to determine how to allocate company resources to achieve the optimal level of safety, productivity, and quality. These are the competing goals that an organization has to balance in order to protect their stakeholders, investments, and assets (Casa.gov, 2014). A thorough investigation can reveal the cost of an accident/incident, satisfy legal obligations by an employer, and assess regulatory compliance (CCOHS, 2011).

The direct costs associated with accidents/incident include medical and indemnity payments, often imposed on employers through insurance claims and lawsuits. In addition to high direct costs, indirect costs can add up to a substantial amount as well when accounting for the number of lost workdays impact on production rate and schedule, training of new employees, legal fees and an increase in insurance premiums. Every year large corporations experience thousands of lost workdays due to workplace injuries, illnesses, and death. Nationwide, these costs add up to well over $150 billion per year. An effective safety program that includes accident/incident investigation process to reduce the risk of injuries in a workplace can save businesses more than four times the amount invested to correct safety issues, through decrease in insurance costs and increase in productivity and quality (OSHA, 2003).

A more proactive way to avoid the costs associated with accidents/incidents is to consider the benefits associated with accident/incident prevention. The benefits can go far beyond the financials discussed above. By investigating and preventing accidents, companies can avoid negative publicity and experience improved morale strengthening the company's credibility and competitiveness through gaining the loyalty of their customers and employees. Furthermore, fewer accidents would mean fewer inspections for compliance and less penalties (Reese, 2011). Perhaps most importantly, accident prevention can save lives and prevent injury.

When an accident or an incident occurs the cost to the organization occurs in two forms: direct and indirect costs. Direct incident cost refers to actual cash outlays that are incurred as a result of the occurrence of the incident. Whereas, indirect cost are costs incurred indirectly. An example of an indirect cost would be the loss of profit due to bad publicity, loss of productivity, the cost of making improvements to the current training protocol, etc. The total incident costs then is the sum of the direct and indirect cost. The following methodology advocates the use of a total incident cost (TIC) worksheet (figure 9.3).

DETERMINING TOTAL INCIDENT COST

First, obtain the average direct incident costs for a given time period. This is accomplished by dividing the sum of the actual cash cost by the number of incidents. Second, the average indirect cost is determined by multiplying the average time (in hours) a department spends addressing the incidents by the average hourly compensation rate. Both direct and indirect costs are then recorded in the TIC worksheet. The sum of the direct and indirect costs constitutes the total incident cost. Incidents here refers to first-aid cases, recordable cases, restricted workday cases, lost-workday cases, permanent disability cases, near-misses and property damage cases; however, what is classified as an incident would be customized to fit the particular organization applied. LaBelle (2000) expounded this further by listing the following steps for completing the direct cost portion of the TIC worksheet:

- Select a reasonable time period to allow for more incidents to be captured for the study.

Direct Costs to the Organization	
Workers' compensation benefits	$________
Healthcare costs	$________
Increases in medical insurance premiums	$________
Auto insurance and liability claims and settlements	$________
Physical and vocational rehabilitation costs	$________
Life insurance and survivor benefits	$________
Group health insurance dependent coverage	$________
Property damage (equipment, products, etc.)	$________
Motor vehicle repair and replacement	$________
EMS costs (ambulance or medivac helicopter)	$________
Vehicle towing, impoundment and inspection fees	$________
Municipality or utility fees for damage to roads, signs or poles	$________
Direct Total $________	
Indirect Costs	
Supervisor's time (rescheduling, making special arrangements)	$________
Fleet manager's time to coordinate vehicle repair, replacement, etc.	$________
Reassignment of personnel to cover for missing employees (less efficient)	$________
Overtime pay (to cover work of missing employees)	$________
Employee replacement	$________
Re-entry and retraining of injured employees	$________
Administrative costs (documentation of injuries, treatment, absences, crash investigation)	$________
Inspection costs	$________
Failure to meet customer requirements resulting in loss of business	$________
Bad publicity, loss of business	$________
Indirect Total $________	
TOTAL $________	

Figure 9.3 Total Incentive Cost Worksheet.

- Select one type of case to review.
- Determine total cash outlay for each category using the TIC worksheet.
- Divide each category cost total by the number of cases for that time period and record the resulting average on the TIC worksheet.

Calculating indirect cost on the other hand is more difficult; however, the following steps are suggested:

- Determine, by category, the average time devoted to an incident.
- Ascertain the hourly compensation rate of those who performed incident-related tasks.
- Determine how much time is devoted (on average) to handling each incident.
- Multiply the average time by the compensation rate to calculate indirect costs for each category (2000).

Adopting this methodology would enable safety professionals to achieve the following:

- Determine and present to top management the true costs of specific incident types.
- Workers compensation and other associated cost would be effectively managed and reduced.
- Drive design, equipment, or process improvements. The information gathered for the TIC worksheet can also help justify upgrades, new programs, and training.
- Enables safety departments to monitor and manage incident cost information (both direct and indirect), in addition to the traditional responsibility of reducing the frequency and severity of incidents and accidents.
- Help in developing a cost/benefit analysis model that will enable the department to show management that the cost of preventing problems is less than the cost of correcting them (proactive approach to safety).
- Explain how the safety department, through effective incident cost management, adds value to the organization.

Using the TIC worksheet would enable safety professionals to articulate, track, and keep a comprehensive record of the true costs of accidents and incidents. In turn, these measures would demonstrate the value-added aspects of safety efforts.

APPROACH TO ACCIDENT/ INCIDENT INVESTIGATION

An effective SMS requires a combination of reactive, proactive, and predictive methods for hazard identification, risk control, and safety data collection. Since accident/incident investigation is data driven, let's consider all three methods that can produce the required data. Reactive methods involve analysis of events that already took place, where the current controls failed to prevent such outcomes. Reports of incidents and accidents through mandatory reporting systems are an example of a reactive data collection method for identifying hazards. Proactive approaches take into consideration current/real-time situations and gathers data through audits, inspections, surveys, and voluntary reporting systems. The goal is to constantly identify new hazards by evaluating existing operations/processes and control them before they lead to an incident. Predictive approaches require extensive data analytics and evaluation of a system (operations/processes and the environment) to identify and mitigate future hazards and possible negative outcomes (ICAO, 2013).

While accident/incident investigation is a reactive response to an unfortunate outcome, it does have a proactive aspect to it, in terms of its intent to find the root cause of the event, identify new hazards, and control them before another similar event takes place. In a more mature SMS, where an investigation is conducted on minor events as well, the contribution to continuous safety improvement can enable this reactive process to be used as a proactive tool. A key aspect of a proactive approach to safety is in communicating the results of an investigation to all levels of personnel as lessons learned in order to create awareness of newly identified hazards and contributing factors and make employees aware of any corrective actions taken to prevent the same incident/accident from happening again (Lynn, 2014).

NONPUNITIVE INVESTIGATIONS: ACCOUNTABILITY VERSUS BLAME

A proactive approach to accident/incident investigation requires appropriate collaboration teams and review committees to be established with consideration for all aspects of operations and all stakeholders to guarantee a blame-free investigation. A nonpunitive organizational culture can ensure that an investigation will be conducted for the purposes of finding systematic/common causes (e.g., malbehaviors that have become part of a system, perhaps encouraged through design, deficiencies in policies/procedures, or the management system) and not special causes (such as individual behavior; Krause & Russell, 1994). Organizations with an effective SMS in place should have their accident/incident investigation process established in writing, clearly capturing the objective of the investigation process as being accident/incident prevention and not determination of liability (ICAO, 2013).

Having a nonpunitive approach to accident/incident reporting and investigation does not mean that employees are not expected to be accountable for their actions within the limits of their roles and responsibilities. Accountability should be balanced and clearly defined for every level of employment. Accident and incident investigation should take accountability into consideration but only for the purposes of resolving any deficiencies in the system, not for the purposes of penalizing parties involved in the event in question (Stolzer & Goglia, 2015). A nonpunitive approach to accident investigation and reporting goes a long way in the workplace by positively impacting every employee. It demonstrates the employer's commitment to providing a safe and healthy environment. Employers can expect an increase in workplace morale and productivity (OSHA, 2015).

INTEGRATING HUMAN FACTORS INTO ACCIDENT/INCIDENT INVESTIGATION

> Human Factors is concerned with the application of what we know about people, their abilities, characteristics, and limitations to the design of equipment they use, environments in which they function, and jobs they perform. —HFES, n.d.

Within any given industry, workplaces involve constant interaction between humans, equipment/systems, and the environment in which they operate. Even in the most automated environments, safety concerns/hazards exist, as you would still need people to monitor the automation, fix the automation when it fails, and control the automation during normal day-to-day operations. One way to look at accidents and incidents is through failures that have taken place leading to a negative outcome. In general, failures can be categorized as System Failures or Operational Failures. System failures refer to the physical equipment failing (display suddenly turns off, a switch breaks, forklift stops moving, etc.). Operational failures refer to failure modes of an operator, the human component (human errors, inadvertent/intentional violations, memory lapse, etc.; HSE, n.d.). System failures are typically traceable and more understood than operational errors. Operational failures, on the other hand, are far more complex because humans are more dynamic and the contributing factors to human behavior are not easily quantified. Thus, the integration of human factors into accident/incident investigation is absolutely required in order to accurately identify and address any contributing factors that impacted human performance.

Accident/incident investigation should be conducted with consideration for human physical and cognitive limitations and an understanding of human behavior. Human reliability is not as quantifiable as system reliability, and human error is inevitable in any given environment where humans operate. Workers are a valuable resource that an organization must protect. Values and attitudes shared by many workers over a long period of time can become systematic, meaning patterns of mal-behaviors may emerge and become *the way things are done* in that workplace. Integration of human factors in accident/incident investigation can help identify such issues and mitigate them through employee motivational techniques, training, and a strong, management-backed safety culture (Reese, 2011).

IT COULD HAPPEN TO YOU

Failure of a system is a feature that is built-in by design. It is not a matter of if, but rather, when, a failure will occur. SMS implements the exercises of preemptive investigation within all phases of design, manufacture, production, operation, and maintenance to identify and manage those inherent risks and hazards. The idea is to apply enough forethought and protection without: (1) breaking the bank and (2) loss of life or property. Accidents will continue to occur, as it is impossible to control for every potential hazard. Perhaps accident and incident investigation, although reactive, is the best way for humans to understand how to think proactively about safety. In this way, the reactive process becomes the impetus for proactivity within the SMS process. By way of illustration, we would like to revisit, the nuclear reactor accident that occurred at the Chernobyl nuclear power facility in Pripyat, Ukraine and also the crash of Japan Airlines Flight 123 at Mt. Osutaka, the deadliest single-aircraft accident in aviation history.

Chernobyl

There have been several accident investigations on the nuclear reactor meltdown that occurred on April 26, 1986. The accident has been blamed on faulty reactor design and inadequate training of operational personnel. It is important to note that at the time of the accident, Ukraine was still part of Russia, and the Cold War was very active between Russia and the United States. When the reactor meltdown took place, there were 47,000 people in Pripyat who required immediate evacuation. However, health dangers were not effectively communicated by authorities, and evacuations were not completed until several days later. Animal evacuations did not occur at all. The resulting human and animal suffering due to nuclear fallout could not be hidden from the world. The effects were so long lasting that the residents of the area are still plagued by adverse medical conditions. Much can be gleaned from the Chernobyl disaster, not only for the nuclear community but for all high-risk industries. Until Chernobyl, the world had not contemplated the need for emergency preparedness and large-scale evacuations of this magnitude.

Japanese Airline Flight 123

The Japanese Airlines Flight 123 was a scheduled domestic flight on August 12, 1985, originating from Tokyo's Hadena Airport outbound to Osaka International Airport, Japan. There were a total of 524 souls onboard including the 15 crew members. Approximately twelve minutes into the flight, prior to reaching cruising altitude, the aircraft suffered an explosive decompression and crashed into the Osutaka Ridge near Mount Osutaka, Japan. Casualties of the crash included all 15 crew members and 505 of the 509 passengers, causing this accident to be the worst single-aircraft accident in aviation history, and the second-deadliest aviation accident after the 1977 Tenerife airport disaster (Kobayashi & Terada, 1985). Japan's Aircraft Accident Investigation Commission officially concluded that the accident was caused by the rupture of the airplane's aft pressure bulkhead. The high-pressure surge into the aft unpressurized portion of the tail caused extensive damage to all four of the airplanes hydraulic lines, and the Auxiliary Power Unit (APU), the rudder, and a large portion of the vertical stabilizer were lost. The complete loss of hydraulic pressure led to a loss of control of the aircraft. Despite severe lateral-directional (dutch-roll) oscillations and pitch, the crew struggled to keep the airplane airborne for approximately 30 minutes while attempting to return to the airport of departure before crashing in mountainous terrain. The root cause of the problem was later traced to an incorrect maintenance procedure adopted by a Boeing repair team that fixed the aft tail section after it was involved in a tail strike incident, which led to damage to the rear pressure bulkhead several years earlier. Further examination revealed that the repair was made using two separate splice plates instead of a single plate as per standard operating procedure. This error resulted in the entire load being transmitted through a center rivet hole causing multiple fatigue cracks from the rivet hole soon after the repairs. This crack propagated, and after a total of 10,000 pressurization cycles, the faulty installation failed leading to the collapse of the bulkhead and rapid decompression. Furthermore, the use of fillet sealant made the repair look normal when viewed by the naked eye during inspection. Many takeaway lessons could be gleaned this accident including the following:

- Aircraft maintenance programs should be developed to reflect consistency with actual aircraft operations.
- Repairs should be accomplished in a manner that preserves the original safety features of the airplane.
- Fatigue properties of complex structural elements should be well understood so that in-service maintenance programs are properly designed (FAA, 2017).

ICAO INVESTIGATION MANAGEMENT SYSTEM

To achieve the objectives and goals, an incident investigation, either major or minor, must be well planned and managed. A qualified individual is required to effectively plan, coordinate, and supervise the intricate and complex processes involved in a major incident investigation. It is this individual's responsibility to assemble an investigation team and ensure synergy and coordinated effort by assigning responsibilities to each group or members of the team while managing the investigation process. It is often the best approach to quickly ascertain the scope and magnitude of the work involved so that adequate manpower, expertise, and resources required for the inquiry can be assembled in an appropriate period of time.

It is often the case that a sizable team of investigators would be required to conduct a major incident investigation. Such a team would likely be comprised of specialist from diverse fields in order to effectively cover all aspects of the investigation. It is best practice for the team to investigate all factors to uncover any underlying systemic causal factors or non-causal discrepancies that might lead to future incidents. A minor incident on the other hand would clearly require less resources and a smaller investigating team to carry out the same functions. A few competent investigators, or alternatively an investigator and a relevant specialist, would suffice. To conduct an investigation in an efficient and expeditious way, ICAO formulated an Investigation Management System "based on a comprehensive plan, checklists, and a method and flow charts to track the progress of the investigation" (OSHA, 2015). The following illustration of an aircraft accident investigation management process draws heavily from ICAO Accident Investigation Manual, Part II:

- The investigation management system breaks the investigation activities into functional events. Each event is identified by a number and a descriptive phase that encapsulates its function. A list of the events in the management system can be found in Appendix B.
- Each functional event is assigned to a group in the team. As many groups as may be required by the complexity of the incident should be formed. Examples of such groups are "Witnesses, Meteorology/Weather, Air Traffic Services, Aircraft Structures, Aircraft Systems, Power plants, Maintenance Records, Survival Factors, Human Performance, Aircraft Performance, and Flight Recorders" (ICAO, 2011). Each group should be headed by a senior investigator and have as many members as is required for it to function effectively. All activities of the group, in relation to its assigned events, are well documented for ease of monitoring. An investigation event task-assignment chart can be found in Appendix C.
- During the course of an investigation, an investigation management flowchart is utilized to provide a continuous update on the progress of the investigation, as well as ensure that the essential sequence of events is followed. The flowchart consists of sets of events that are executed sequentially by the investigators.
- Every investigation management system event has a checklist. The checklist is often customized to the peculiar nature or circumstances surrounding the incidents. It is a tool that grants the investigator-in-charge (IIC) the chance to effectively order the conduct of the investigation, monitor the

output of the various investigators, and provide guidance to first time participants. The detailed Investigation Management System event checklist can be found in Part II of ICAO Manual of Aircraft Accident and Incident Investigation (ICAO, 2015b).

- A daily progress meeting between all parties involved in the investigation also forms part of the investigation management process. The meeting is hosted by the IIC, and it offers the chance for the reports and findings of other groups to be shared and discussed. The brief report to be presented by each group chairpersons follows the following format: (1) What we did today, (2) what we found today, (3) what we plan to do tomorrow, and (4) any comments, questions, or suggestions.

The investigation management system is a potent element in the toolkit of a potential IIC, or group chairperson, likely to be involved in a major incident investigation; hence, it is expedient that both should be quite familiar with the system before attempting to use it in the field. The effectiveness of the system is dependent on the diligent employment of the flow charts and the checklist. This offers the IIC an effective means of monitoring the progress of the investigation, while the group chairpersons have a tool to organize the work of their groups.

INCIDENT INVESTIGATION PROCESS

Incident investigation is a potent and effective tool in enhancing safety in any organization. Its effectiveness, however, depends upon the approach and the process adopted to conduct the analysis. The set objective of any investigation should be established at the early stages and should be focused on determining the direct and indirect cause(s) of the incident, evaluate response actions, identify corrective actions, and proffer pertinent recommendations to avert any recurrence of a similar incident in the future. Any incident involving injuries to persons or properties will likely involve the assignment of blame and possible legal proceedings; therefore, instances of this nature should be handled with the utmost care. The investigators run the risk of losing the cooperation of vital persons involved if employees are worried about being blamed for the occurrence (Ashbrook, 2014). Typically, investigations are conducted by safety officers, safety committee members, or supervisors. Truly, anyone can carry out an incident investigation. However, the individual selected to carry out an incident investigation must be a good communicator who engenders trust and is regarded as competent by all concerned. In terms of numbers, a large team of investigators is usually required when it involves a major incident or a complex situation. A non-major incident, on the other hand, might require one investigator and possibly a few assistants with complementary technical expertise.

ICAO (2003), defines accident investigation as "a systematic process whereby all of the possible causes of an adverse event are evaluated and eliminated until the remaining causes are identified as applicable to that investigation." ICAO stresses the need for investigators to hold judgment and keep an open mind so as to not fixate on readily identifiable causal factors and inadvertently overlook other key influences. Since incidents and accidents are rare occurrences, minor incident investigations should be used as learning tools to offer investigators the opportunity to train and acquire needed skills and build capacity. ICAO, in its manual on aircraft accident and incident investigation, articulated the investigation process as consisting of three phases: collection of the data, analysis of the data, and the presentation of findings. The following surmises the key components of that process.

COLLECTION OF THE DATA

The first task during the commencement of an investigation is defining and collecting data relevant to the incident. To help guide the data collection, it is helpful to keep the following issues in focus: (1) what activity was ongoing when the incident occurred, (2) causes of the incident, (3) contributing factors, (4) response actions, (5) preventive measures, (6) mitigating measures, and (7) corrective actions against future occurrence (Ashbrook, 2014). For an aircraft accident investigation, data collection is a continuous process throughout the life of the investigation as more information comes to light. Hence, the fragility and perishability of data is a critical concern and should be addressed. The pertinent data to be collected should include:

- *Accident particulars.* Reference data such as date, location, time, cruising altitude, departure point, ETAs and ETDs, destination, and enroute stops should be collected to aid the analysis and interpretation of other vital information.
- *Meteorological particulars.* This refers to actual or forecasted weather conditions that may have impacted the flight conditions or aircraft performance leading to the accident. Such data would include atmospheric conditions, visual illusions, windshear, and issues that may have affected the aircraft's takeoff or landing profiles.
- *Technical particulars.* The sources for technical data vary from onboard recording devices, manufacturing and maintenance records, results of component analysis, to review of past incidents from relevant databases. A vital benefit of technical data is the possibility of recreating and simulating the likely chain of events leading to the accident.

- *Human factors particulars.* This set of data is often the most difficult to obtain due to the limitations of the human element. Humans are liable to forget and distort facts and are vulnerable to overt and subjective influences. Additionally, the individuals involved in fatal accidents are inaccessible and few witnesses would be available to give an account of what the crew was experiencing leading to the accident. Witnesses, when available, need to be managed to ensure an accurate and uncontaminated rendition of events.

DATA ANALYSIS

Data analysis occurs concurrently with data collection. Rarely do analysis results reveal an isolated cause with clear and specific resolutions. In most cases, analysis reveals more questions and hence the need for further information and data collection.

PRESENTATION OF FINDINGS

The findings and final report must reveal, in detail, what happened, how it happened, and why it happened. It is the required stimulus to preventing future occurrences. Thus, the final report, including safety recommendations derived from the findings of the investigation, is released in a timely fashion. This also ensures that all parties involved are kept informed of the development of the investigation and any safety deficiencies that may have been uncovered, as well as any corrective action that will take place to mitigate the uncovered risks.

ACCIDENT PREVENTIVE MEASURES

One of the key benefits of conducting incident investigation is to prevent the recurrence of similar incidents. The investigation of incidents is an effective method of preventing accidents. This is achieved by the identification of corrective actions often derived from the analysis of accident and incident data. Numerous incidents, without any associated injuries or damages, do occur regularly, and they offer abundant hazard information. A successful and effective accident preventive measure should employ the following components:

- An incident reporting system should be established for the collection of safety-related deficiencies and lapses. It should comprise both a mandatory and a voluntary incident reporting system. The essence of the voluntary reporting system is to facilitate the collection of safety information that may not be captured by the mandatory system; hence, it should be nonpunitive and designed to offer protection to apprehensive sources. A nonpunitive and benign setting is a necessary precursor to a voluntary incident reporting system that would motivate people to report hazards, deficiencies, and discrepancies in which they were involved or observed. De-identification is one way of ensuring confidentiality in such a system, which is why voluntary reporting systems have been more successful than mandatory systems in capturing human-factor-related information (ICAO, 1984).

- A database system should be established to facilitate the gathering, collation, analysis, and exchange of aggregate safety information. The database should have a harmonized format to aid data sharing, while such information sharing mechanism should be networked and collaborative.
- Collated information from existing incident and accident reporting system, accident investigation reports, safety audits, databases, and so forth, should be analyzed with regularity in order to identify pertinent preventive actions.
- A safety data and information sharing network should be instituted to facilitate the free flow of potential and actual safety discrepancies and deficiencies that may be of value to other stakeholders or partners in the aviation system. Harmonized taxonomies and formats enormously aid information sharing among diverse industry stakeholders (ICAO, 2001).

FOLLOW-UP ON MULTIPLE LEVELS

Once an accident investigation is conducted, the organization may be left with be numerous contributing factors, lessons learned, and corrective actions. To extract the maximum benefit from the results of investigative reports, it is necessary to follow-up with these investigation results on multiple levels. The corrective actions and lessons learned from the event must not only be realized and documented, but they must also be clearly communicated to the relevant parties within the organization. Those affected by changes to the organization should be informed of these changes in a timely manner.

PREDICTIVE ANALYSIS—TURNING HINDSIGHT INTO FORESIGHT

Predictive safety analysis involves developing a set of numerical modeling tools that help to predict the safety of a system based upon the traditional risk assessment combined with data gleaned from accident and incident investigations. The data collected from the investigation, as well as standard operating data, is used to create an algorithm to assess the level of safety within the organization. Similar algorithms have been developed and are in use by Southwest Airlines and Embry-Riddle Aeronautical University. The idea is to use predictive data from known accident conditions, including human factors components, to better monitor and predict the factors that contribute most greatly to an accident or incident occurring.

REFERENCES

Ashbrook, P. C. (2014). Incident investigations. *Journal of Chemical Health and Safety*, *21*(4), 47. http://doi.org/10.1016/j.jchas.2014.05.009.

Canadian Centre for Occupational Health and Safety. (2011). Answers Fact Sheets. *Occupational Health and Safety*. Retrieved from https://www.ccohs.ca/oshanswers/occup_workplace/.

CASA.gov. (2014). SMS for aviation: A practical guide. *Safety Assurance*. Retrieved from www.casa.gov.au.

Deputy Press Secretary Larry Speakes. *"U.S. Reaction to the Chernobyl Explosion,"* Making the History of 1989, Item #174. Retrieved from https://chnm.gmu.edu/1989/items/show/174.

e-CFR. (2016). 830.2 Definitions. *Electronic Code of Federal Regulations*. Retrieved from www.ecfr.gov.

Federal Aviation Administration. (2015a). *What's Happening with SMS. Keeping Flight Standards and Industry Informed. SMS Program Office*. Retrieved from https://www.faa.gov.

Federal Aviation Administration. (2015b). Safety Management Systems for aviation service providers. *AC 120-92B*. Retrieved from www.faa.gov.

Federal Aviation Administration. (2017). Japan Airlines Flight 123, Boeing 747-SR100, JA8119. *Lessons Learned*. Retrieved from http://lessonslearned.faa. gov/ll_main.cfm?TabID= 4&LLID=16&LLTypeID=2.

Geigle Safety Group. (2016). Guidelines for employers to reduce motor vehicle crashes. *OshAcademy*. Retrieved from https://www.oshatrain.org/notes/4dnotes02.html.

Health and Safety Executive (HSE). (n.d.). HSE human factors briefing. *Humans and Risk*. Retrieved from www.hse.gov.uk.

Human Factors and Ergonomics Society (HFES). (n.d.). Definition of human factors and ergonomics. *Educational Resources*. Retrieved from www.hfes.org.

ICAO. (1984). *Accident Prevention Manual (Doc 9422)*. Retrieved from https://www.scribd.com/document/357353625/Doc-9422-Accident-Prevention-Manual

ICAO. (2001). Aircraft Accident and Incident Investigation (Annex 13). Retrived from https://store.icao.int/collections/annexes.

ICAO. (2003). *Doc9756 Manual of Aircraft Accident and Incident Investigation Manual of Aircraft Accident and Incident Investigation – Part III (Investigation)* (Vol. IV). Retrieved from https://www.public.navy.mil/navsafecen/Documents/aviation/investigations/ICAO_Inv_%20Man_Pt3.pdf.

ICAO. (2011). *Manual on Accident and Incident Investigation Policies and Procedures Doc 9962 AN/482*. Retrieved from http://aaib.gov.mn/wp-content/uploads/2018/12/Doc9962-Manual-On-Accident-And-Incident-Investigation-Policies-And-Procedures.pdf.

ICAO. (2013). *Safety Management Manual (SMS)*. Third Edition. Doc 9859. Retrieved from http://www.icao.int/safety/.

ICAO. (2015a). *Manual of Aircraft Accident and Incident Investigation Part I – Organization and Planning*. Retrieved from https://www.skybrary.aero/bookshelf/books/3282.pdf.

ICAO. (2015b). *Manual of Aircraft Accident and Incident Investigation Part II – Procedures and Checklists*. Retrieved from https://store.icao.int/products/manual-of-aircraft-accident-and-incident-investigation-part-ii-procedures-and-checklists-doc-9756p2

Kobayashi, H., & Terada, H. (1985). Crash of Japan Airlines B-747 at Mt. Osutaka. *Failure Knowledge Database*. Retrieved from http://www.sozogaku.com/fkd/en/cfen/CB1071008.html.

Krause, R. T., & Russel, R. L. (1994). The behavior-based approach to proactive accident investigation. *Professional Safety, 39*(3), 22. Retrieved from http://search.proquest.com.

LaBelle, J. E. (2000). What do accidents truly cost? *Professional Safety, 45*(4), 38–42.

Lynn, G. D. (2014). Incident investigations: Four critical stages. *Professional Safety, 59*(9), 53.

OSHA. (n.d.). Safety and health topics. *Incident investigation*. Retrieved from www.osha.gov/.

OSHA. (2003). Fall protection-it's a snap! *Safety Pays*. Retrieved from www.osha.gov.

OSHA. (2009). Accident investigation basics: How to do workplace accident investigation. *Incident (Accident) Investigations: A Guide for Employers*. Retrieved from https://www.osha.gov/.

OSHA. (2015). A systems approach to help prevent injuries and illnesses. *Incident (Accident) Investigations: A Guide for Employers*. Retrieved from www.osha.gov.

Reese, C. D. (2011). *Accident/Incident Prevention Techniques* (2nd ed.). Baton Rouge, LA: CRC Press.

Stolzer, A., & Goglia, J. (2015). *Safety Management Systems in Aviation* (2nd ed.). Burlington, VT: Ashgate.

Stpaulssteinerschool.org. (2013). Accident/incident definition. *Definitions*. Retrieved from www.stpaulssteiner school.org.

Chapter 10

Safety Management Systems for Rotorcraft Manufacturers

Isaac Munene

This chapter focuses on the task of developing an integrated or comprehensive enterprise SMS for a generic rotorcraft manufacturer. It is intended to provide some insight into the complexities and challenges a rotorcraft manufacturer will experience in meeting the requirements of an SMS when economic (business) and regulatory requirements are considered. This chapter can be considered as an example or expansion on to the work of Yantiss (2011), *Implementation of safety management systems in aviation.* This chapter reviews the existing regulations or guidance on SMSs, describes the likely global presence of a generic rotorcraft manufacturer based the current industry players' global locations, defines the structure of an integrated SMS with its relationships and dependencies, identifies the challenges to an integrated SMS, and finally, provides recommendations for achieving an integrated SMS that meets the needs of the organization.

SAFETY MANAGEMENT SYSTEMS

The ICAO, as the leader in global aviation, develops the principles and techniques by which air navigation is conducted and fosters the planning and development of international air transport to ensure it is safe. As a result of a mandate to implement an SMS, also referred to as the State Safety Programs (SSPs) for their respective aviation industries, the first *System Safety Manual (SMM) Doc 9859* (ICAO, 2006) was developed as guidance material. By then, Transport Canada had already led the way by adopting a SMS in 2005, and the United States and Europe have since followed (Stolzer, Halford, & Goglia, 2008). The individual states and regulatory bodies have developed additional regulations to facilitate the implementation of the SMSs by the industry within their areas of jurisdiction.

The FAA, in following the ICAO model, has developed and proposed SMS requirements for the aircraft original equipment (airframe and engine) manufacturers, maintainers, and operators to implement as a means of managing safety with the necessary organizational structures, accountabilities, policies, and procedures in place. The SMS, thus, becomes an organization-wide comprehensive and preventive approach to managing safety that encompasses safety policy, formal methods of identifying hazards, mitigating risk, and promoting a positive safety culture (FAA, 2012; ICAO, 2009). The FAA acknowledges the significant efforts to develop and integrate a SMS, and therefore, the FAA provides guidance to the industry in meeting the requirements.

To maximize the chances of organizations successfully implementing a SMS in an efficient manner, the FAA has been conducting voluntary SMS pilot studies with a few service providers in order to determine the scope and detail of guidance to be provided to the industry at large. The FAA's Flight Standards, Airports, and Aviation Safety organizations have or will be using the voluntary pilot projects to: (1) develop implementation strategies, (2) develop oversight interfaces, and (3) gain experience for the FAA and service providers for the development of a voluntary SMS (FAA, 2015). The SMS pilot projects have been initiated by each organization with the support of the respective FAA offices. The Flight Standards organization is responsible for the Operators and Product/Service Providers program (AVS SMS); Aircraft Certification Service (AIR), the design and manufacturing (D&M SMS), and FAA Airports, the airports' program (FAA, 2015). The AVS SMS covers flight operations and maintenance activities within the organization. A comprehensive SMS integrates all SMS thereby covering all aviation safety related activities through an aircraft's lifecycle.

The desire to develop a SMS presents a unique challenge for any rotorcraft manufacturer when all three areas covered by the pilot projects are considered. The company, as part of its design and manufacturing function, will perform flight tests with the production aircraft to ensure all regulatory and certification requirements are met prior to the aircraft's introduction into service. The organization may also operate a flight department from within its own airport, or one shared with other operators, for its corporate or executive travel needs. Maintenance of the aircraft is performed in a consistent manner to ensure the readiness for safe operations. To address all critical areas of safety identified, an organization should consider being an active participant in all three projects as a good start toward an integrated SMS.

An integrated SMS is one that is capable of meeting the needs and goals of the different functional groups as well as the entire organization as a whole. Rotorcraft manufacturers, like many other companies, are experiencing the effects of globalization and have made strategic decisions to have subsidiaries, operating business units, affiliates, or joint ventures in different regions of the world in order to expand their market reach. According to PricewaterhouseCoopers (2013), the leading aircraft manufacturers, is making concerted efforts to expand their global footprint in order to win a share of the local markets, as many countries, such as China and Russia, are requiring manufacturers to partner with local companies to design, build, and manufacture the aircraft that will serve that market. Honeywell (2015) estimates a demand increase of approximately 5,500 new civilian helicopter deliveries between 2015 and 2019. This may continue to fuel the push for foreign operations. With the increased demand for aircraft, a need for trained and experienced technicians, pilots, and mechanics will exist in all regions to support operators.

As the aircraft safety regulations differ from country to country or region to region, so do the requirements for the SMS adopted by the country's SSP. To be financially prudent and effective, organizations are looking at the likelihood of having an integrated or enterprise SMS akin to the harmonization of the SMS regulations around the globe. An enterprise SMS allows the management team across all levels to have oversight on safety for all the locations and access the data easily. To identify the challenges or barriers to the development of an integrated SMS, a review of the requirements of the SMS voluntary projects (AVS SMS and D&M SMS) as described in AFS-900-002-G201 (FAA, 2014a), and the organization's possible global footprint, as suggested by the presence of other rotorcraft manufacturers, was made.

ROTORCRAFT INDUSTRY

Rotorcraft manufacturing is a global industry featuring key players and products ranging from light, single-pilot aircraft to multicrew and multi-engine aircraft. The Strategy Guide-2015-World's 6 Leading Helicopter & Rotorcraft Manufacturers—Key Strategies & Plans - Airbus Helicopters, AgustaWestland, Bell Helicopter, Boeing, Russian Helicopters, Sikorsky report (Research & Markets, 2014) identifies the principal manufacturers in the industry. To compete in the global market, these manufacturers have expanded from their domestic operations to regions considered as emerging markets. With this expansion, each manufacturer is required to adhere not only to the local business practices, laws, and standards, but also the global regulations, such as the development of a SMS as required by ICAO.

To understand the complexity of developing an integrated SMS, the probable global footprint of a generic rotorcraft manufacturer is generated by mapping the regions of operation of the six leading manufacturers (as identified on their respective websites) and the associated regulatory body. For a generic operator, following the ICAO SMM model may be sufficient to meet the SMS requirements in the countries of operations except in countries where the SSP has add-on requirements. At any given instance, a major rotorcraft manufacturer may be subject to the SMS requirements of one or more of the following aviation regulators: (1) the FAA, (2) the EASA, (3) Brazil's Agência Nacional de Aviação Civil (ANAC), (4) Transport Canada Civil Aviation (TCCA), (5) United Kingdom's Civil Aviation Authority (CAA), (6) Australia CASA, and (7) the Civil Aviation Administration of China (CAAC). The unique requirements of each SSP should be met under an integrated SMS developed by an organization's subject matter experts.

IMPLEMENTATION OF AN INTEGRATED SMS FOR A ROTORCRAFT MANUFACTURER

ICAO defines a SMS as "an organized approach to managing safety, including the necessary organizational structures, accountabilities, policies and procedures" (ICAO, 2013, xii). The goal of the SMS is to continuously improve safety by identifying hazards, collecting and analyzing data, and assessing the safety risks in a proactive manner. For a rotorcraft manufacturer to achieve this type of continuous improvements in safety, the integration of two of its organizational management systems—the QMS and the SMS (ICAO, 2013, 2.9.1)—will need to be accomplished. The QMS, described as the organization, its people, and operational processes that are dedicated to meeting customer requirements for the delivered products and services, should foster continuous improvement as implied by the quality principles of the International Standards Organization (2012). The SMS will be comprised of the AVS SMS, D&M SMS, and where applicable, the Airport SMS, combined.

For efficiency and ease of integration, each SMS can be designed with existing organizational processes and systems in mind by tailoring the flexible regulatory framework to the organization. The SMS design should support its five key processes: hazard identification, occurrence reporting, risk management, performance measurement, and quality safety assurance. Approaching the SMS from Transport Canada's perspective can create a gradual shift in regulatory philosophy from one that emphasizes continuous compliance inspections to one that focuses on the operator's performance on the twelve elements of the SMS (Yantiss, 2011). These twelve elements are embedded in the four component structure defined in the ICAO SMM (2009) as shown in table 10.1. These elements provide some visibility to the requirements of

Table 10.1 ICAO SMS Components and Elements

1. Safety Policy and Objectives 1.1 Management commitment and responsibility 1.2 Safety accountabilities 1.3 Appointment of key safety personnel 1.4 Coordination of emergency response planning 1.5 SMS documentation	2. Safety Risk Management 2.1 Hazard identification 2.2 Safety risk assessment and mitigation
3. Safety Assurance 3.1 Safety performance monitoring and measurement 3.2 The management of change 3.3 Continuous improvement of the SMS	4. Safety Promotion 4.1 Training and education 4.2 Safety communication.

Note. Adapted from "SMS Implementation," by Yantiss, B., 2011, In *Implementing safety management systems in aviation, p. 180,* Stolzer, A., Halford, C., & Goglia, J. (Eds.). Burlington, VT: Ashgate.

an organizational SMS. The elements can be further broken down to accommodate the size and unique needs of each particular organization.

The goal of implementing an integrated SMS is to ensure the program meets the standards set forth by ICAO, the SSPs of the country of registration and operation, and the overall organization. Having one SMS allows the enterprise business needs, including regulatory compliance, quality assurance, and safety accountability, to be met from one common system. With ICAO Member States being at various stages of SMS adoption and implementation, rotorcraft manufacturers may experience high levels of frustration with their tardiness in defining the regulatory requirements that will directly impact the organization's SMS (Stolzer, Halford, & Goglia, 2011). This frustration may be tempered by participating in the FAA pilot projects to develop a baseline system and modifying as needed to meet ICAO requirements.

To achieve a comprehensive SMS, the existing and newly formulated SMS modules should be seamlessly integrated with the organization's enterprise management systems. In his research on SMS implementation, Yantiss (2011) emphasizes that large organizations are likely to have thirty plus enterprise management systems and provides examples of those that are fundamental to an airline. Together with some of those systems, the systems listed in table 10.2 can be considered applicable to a generic rotorcraft manufacturer at various locations of operation.

The development of an integrated SMS in support of a comprehensive approach to safety management within an organization can quickly become a complex endeavor. Figure 10.1 is a graphical representation of a comprehensive SMS integrated within the larger enterprise management system with the relationships and dependencies of the system identified.

Table 10.2 Enterprise Management Systems for a Generic Rotorcraft Manufacturer

a) an organizational management system (corporate governance)
b) a SMS

1. an aviation safety SMS (AV SMS)
2. a manufacturing SMS (D&M SMS)
3. an airport SMS (APT SMS)

c) an engineering design and tools management system (EDTMS)
d) an environmental, health and safety management system (EHS)
e) a data and security management system (SeM)
f) a documentation management system
g) a flight operations management system (FOMS)
h) an aircraft maintenance management system (AMMS)
i) a dispatch management system
j) a ground handling management system
k) a supplier management system
l) a marketing management system (MMS)
m) a personnel management system
n) a production management system
o) a customer support management system
p) an aircraft delivery management system
q) a quality management system (QMS)
r) a fielded fleet data management system
s) a warranty management system
t) a flight safety data management system
u) a product reliability management system (RMS)

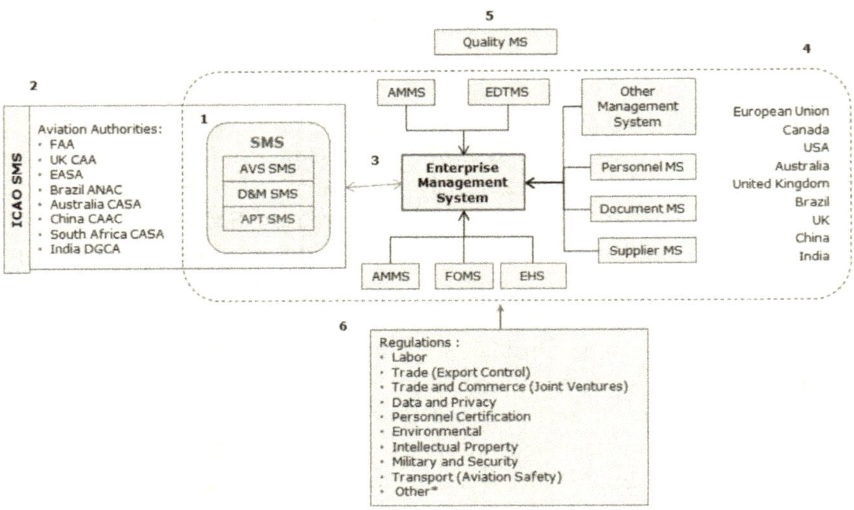

Figure 10.1 Relationships and Dependencies of an Integrated SMS.

The relationships and dependencies identified with numbers are as follows:

1. The SMSs for aviation safety, design and manufacturing, and airport (if required) are rolled into one common system.
2. Each SMS draws its requirements and standards from the SSP and by the ICAO SMS guidelines. The SSP, depending on the country's experience with SMS, may have more stringent requirements beyond those recommended by ICAO.
3. Data exchange occurs within the enterprise management system with the SMS receiving information required to undertake safety assessments and investigations.
4. With the integration of each location's SMS and enterprise management system with similar systems in other locations, the SMS can be considered a comprehensive one for the entire organization. Regulations within the identified regions will influence the success of the SMS.
5. The quality of the data, processes, and systems has to be monitored. Quality assurance across the entire organization ensures that the SMS supports the achievement of the goals set on safety performance.
6. The interactions depicted in figure 2 indicate the introduction of dependencies resulting from the regulations within the region. Data sharing, for example, may be enhanced or restricted by the regulations originating from the individual countries. These dependencies need to be addressed for it to be successful and effective. Each regional or state safety plan provides additional requirements which can create challenges in harmonization with ICAO SMS requirements as observed with the on-going global air traffic management system modernization (FAA Air Traffic Organization, 2014).

CHALLENGES (CONSIDERATIONS) TO AN INTEGRATED SMS

The considerations or challenges that need to be addressed in the development and implementation of the proposed integrated SMS are identified through the detailed review of the four components of safety and existing, or proposed, SMS guidance and regulations.

Implementation costs. The implementation of a new SMS or the integration of a standalone SMS into one global rotorcraft manufacturer, like any other technology investment, can come with significant costs. These costs are associated with the setting up of the systems, tools, and processes required to support the SMS and its integration into the enterprise management system. The effective implementation of a SMS can be time consuming, expensive, and is likely to impact team and organizational performance in the initial implementation stages. For the effective implementation of an integrated SMS, a significant investment of time and money will be required for training, technological modifications, support, monitoring, and performance evaluation (Repenning & Sterman, 2002; Klein & Knight, 2005).

Management support and commitment. The successful implementation of a SMS requires a high level of support and commitment by management to ensure the set safety goals are being achieved. Management should ensure that safety receives the same commitment and effort as other high-priority organizational goals. Management should ensure workers are trained and well-informed of the organization's SMS, as all employees play a significant role its success through the communication of safety responsibilities, developing continuous improvement initiatives, and promoting accountability for safety responsibilities (Friend & Kohn, 2010). The acceptance of safety policies and the building of a just culture within an organization requires upper level management's unwavering support and commitment.

Safety policy standardization. Aviation organizations must have senior management committed to safety by developing and communicating safety policies that allocate responsibilities, thus holding employed accountable for safety performance goals. The formulation and documentation of the safety policy may be required to meet standards defined by the aviation regulator (Transport Canada, 2001). The number of regulators that interact with a global rotorcraft manufacturer differs by the areas of operations. A clear declaration on safety commitment and objectives should meet the requirements of each regulator and the intent of ICAO's guidance. Each SMS will eventually be a combination of the aviation safety, airport, and manufacturing SMS to conform to the regulators' requirements while also meeting ICAO standards. Harmonization of safety policy across the different business locations is important if business metrics, such as the return on safety investment, are to be tracked.

Segmented international collaboration. ICAO, the FAA, and other safety authorities have realized the need for collaboration to ensure the individual states will benefit. ICAO requires the full integration of a SMS for the management of safety risk in air operations, maintenance, air traffic services, aerodromes, flight training, and the design and production of aircraft. ICAO has mandated that states establish SSPs for their civil aviation systems. Failure to harmonize the SMS and SSP requirements within each state and across the globe can put a heavy burden, both in terms of monetary loss and loss of resources such as time, on a rotorcraft manufacturer with multiple locations across the globe. Segmentation may result if collaboration is not emphasized early enough. Segmentation was a major issue during the modernization of air traffic management (ATM) systems, specifically in other regions of the world that tend to lag behind in terms of technological resources. This poses a significant problem when a nation is technologically unprepared for the high air traffic they experience regularly, such as the case of the Gulf States (Turner, 2013).

Transport laws and safety culture. An organization's culture is defined by the decisions and actions of employees that paint a picture of the values held by the organization. An organization's culture is influenced and shaped by the diverse groups that exist within it (Alvesson, 2012). For a strong safety culture to exist, a reporting and just culture must be cultivated but can be hindered by the use of punitive measures. In situations where accidents occur, the fear of judicial action to apportion criminal or civil liability may inadvertently have a negative impact on the safety culture within the organization. In countries allowing judicial action, the fear of providing safety information that may be used against an organization in a criminal prosecution may prevent the organization from supplying voluntary information for safety improvements. Understanding the impact of the legal system on employees' perception of the organization's safety culture is important for the overall success of the SMS. Safety professionals can then use this baseline information to develop a strategic plan to reshape the organization's safety culture for the better.

Trade and commerce practices. An expansion of rotorcraft manufacturers into emerging markets has been achieved through the establishment of joint ventures and new facilities to serve as customization centers, manufacturing or research centers, and flight training centers. A major concern within the industry is managing safety without violating commerce or trade laws that forbid the transfer (export) of strategically important technology, products, services, and information to foreign governments or nationals due to strict national security and foreign policies. Violation of the International Traffic in Arms Regulations or Export Administration can result in serious administrative, civil, and criminal consequences (Blount Jr., 2007). It is critical for rotorcraft manufacturers to tailor their SMS, design, and manufacturing protocol to pay respect to the local trade and commerce regulations within their region of business.

Aviation labor laws and practices. Every organization is required to uphold domestic and international aviation labor regulations and best practices at each location of operations. These regulations are put in place to protect workers and organizations from exploitation and unfair practices. As part of safety management, accident investigations performed by the organization require safety personnel to be cognizant of labor regulations and internationally accepted standards. Research has shown that airlines choosing to outsource the maintenance of their aircraft in foreign regions of operation have experienced issues in training, quality control, skill training and experience, language skills, regulatory oversight, and lack of drug and alcohol testing program (Franck, Lewis, & Udis, 2012; Tang & Elias, 2012). Best practices must be in compliance with the necessary regulations; though, this may introduce new conflict. For example, since 2012 the FAA has sought to expand the alcohol and drug testing requirements for persons performing safety-sensitive maintenance functions on commercial air carrier aircraft by having specimens sent back to the United States. The conflict between the FAA proposal, and the local Civil Aviation Authority's current regulatory policy that does not require drug testing, may put the two at odds (Director of Maintenance, 2014). An integrated SMS should be robust enough to accommodate the differences in regulations at each location, while still supporting the overall goal of improving aviation safety performance.

Data security and privacy. An integrated SMS requires extensive sharing of data across organizations and locations. A rotorcraft manufacturer has to assess the regulations at every region of operation and determine if an integrated SMS is plausible. Recent incidents of cyber security attacks have highlighted how the sources and types of threats to computer systems, electronic networks, stored data and information have evolved and continued to change (Gerber & von Solms, 2001). For a reporting culture to be fostered, a means of reporting safety issues anonymously and securely across the enterprise, regardless of location, should be made available and accessible for employee use. To achieve the highest level of asset protection and information security, the organization will need to invest in the appropriate tools and technology to maintain the privacy of the organization's data; policies and procedures should be adapted to ensure employees are handling, storing, and disposing of confidential data appropriately (Tsiakis & Stephanides, 2005; Eloff & von Solms, 2000).

Intellectual property. A manufacturer's competitiveness is based on the value of its intellectual property (IP). The protection of an organization's IP varies across countries (Budde-Sung, 2013). When a design and manufacturing SMS is implemented, all hazards and risks associated with the aircraft and its components and systems will analyzed during the development and sustainment periods. These analyses may extend to the manufacturer's supply chain where critical intellectual property may be susceptible to exposure within an integrated SMS. IP may also be susceptible to disclosure when an aviation regulator changes policy or regulations. As an example, the FAA issued a policy statement "Type Design Approval Holder Inappropriate Restrictions on the Use and Availability of Instructions for Continued Airworthiness" to address the inappropriate restriction on the distribution of the information to third party maintenance organizations and operators (FAA, 2012). Even though the intent of the policy was clear, it was met with resistance, as the policy was seen as a possible way of reducing an OEM's competitiveness. Depending on the individual SSP, regulators may implement additional SMS policies that may inadvertently impact the organization's IP through disclosure. Understanding such limitations with respect to IP is critical.

RECOMMENDATIONS AND CONCLUSIONS FOR THE INTEGRATED SMS

The goal of SMS is to address safety issues within an organization in a holistic manner. With manufacturing, aviation safety and airport SMS being introduced by ICAO for all member states, it is necessary to have them integrated into the enterprise's organizational management system. For a global manufacturer, this includes all business units or operational facilities located all over the globe. To support an integrated SMS, ICAO, and FAA should consider adopting the same approach taken for the harmonization of the air traffic management systems currently in development, namely NextGen, SESAR, and Japan's Collaborative Action for Renovation of Air Transport Systems, to address air traffic needs in the different regions of the world (Government Accountability Office, 2011; European Organization for the Safety of Air Navigation, EUROCONTROL, 2012). Adopting Transport Canada's approach to the aviation safety SMS and that of the FAA's to the design and manufacturing SMS could harmonize the requirements across all member states.

In 2009 the EASA, ICAO, FAA, and Transport Canada held a meeting to discuss the potential for SMS and SSP collaboration in which the participants agreed that there was substantial value in collaborating. The Safety Management International Collaboration Group (SM ICG) was established with a goal of meeting semi-annually. The SM ICG membership currently includes the major aviation regulators across the globe and interacts with aviation stakeholder groups including industry and safety teams (FAA, 2014b). This collaboration effort should be sustained to ensure all opportunities for collaboration are exploited and the industry can see the benefit of having an integrated SMS. The acceptance or recognition of the approval given to an organization's SMS in one location, for example the United States or Canada, should be considered adequate to meet the SMS standards across other regions. It also eases the financial burden an organization is likely to bear over a long term.

Integrating the organization's SMS involves a significant investment in resources in terms of time and finances. For this to be successful, the SMS effort requires the support and commitment of the management. These two challenges, of resources and commitment, can be addressed by gaining the chief executive and his or her management team's support through both language and actions. Aligning the SMS objectives with the overall business objectives and identifying suitable metrics for performance criteria is necessary. With the goal of a business being the provision of goods and services at a profit, its recommended that a return on investment should be provided to management to show that the SMS improves the bottom line of the company over the long term (Friend, 2011).

Developing and integrating a comprehensive SMS may seem like a daunting task to many manufacturers. Thus, it is recommended that large manufacturers of airframes, avionics, and engines volunteer to participate in the FAA's pilot projects to receive assistance in developing SMS implementation strategies, developing oversight interfaces, and receive support from the local FAA office simply at the price of volunteering (and committing) to integrate a SMS (FAA, 2015). An integrated and comprehensive SMS will facilitate the sharing of safety information across regions and groups providing visibility to new hazards or risks allowing for effective controls to be incorporated. Individual lessons learned on integration can be shared with the industry and, in the process, facilitate other industries with global operations to follow suit.

A rotorcraft manufacturer will also face challenges when developing an integrated and comprehensive SMS that meets the regulations adopted by ICAO's individual member states for the management of labor, transport, data management, and intellectual property. The harmonization of aviation safety regulation by ICAO across regions should be a guiding principle for handling differences in other areas that impact aviation safety management. To begin this effort, the criminalization of accidents that may occur during training operations, flight test, corporate flights, and as a result of maintenance errors should be addressed. All members should agree to adhere to the common standard for aviation safety where the goal is to determine the causal factors to identify underlying risks and hazards in order to optimize safety improvements.

REFERENCES

Alvesson, M. (2012). *Understanding Organizational Culture*. Thousand Oaks, CA: Sage.

Blount Jr., P. J. (2007). The ITAR treaty and its implications for U.S. Space Exploration policy and the commercial space industry. *Journal of Air Law and Commerce, 73*, 705–722.

Budde-Sung, A. (2013). The invisible meets the intangible: Culture's impact on intellectual property protection. *Journal of Business Ethics, 117*(2), 345–359.

Director of Maintenance. (June 2014). *FAA Seeks Overseas Drug & Alcohol Testing Information*. Retrieved from http://www.dommagazine.com/article/faa-seeks-overseas-drug-alcohol-testing-information.

Eloff, M. M. & von Solms, S. H. (2000). Information security management: A hierarchical framework for various approaches. *Computers and Security*, *19*, 243–256.

European Organization for the Safety of Air Navigation. (2012). *Partnering to Deliver Global Interoperability*. Retrieved from https://www.eurocontrol.int/publications/partnering-deliver-global-interoperability-factsheet.

FAA. (2012). *Type Design Approval Holder Inappropriate Restrictions on the Use and Availability of Instructions for Continued Airworthiness*. Retrieved from http://www.pmamarpa.com/gvt/PS-AIR-21_50-01ICAMemo.pdf.

FAA. (2014a). *Safety Management Systems Voluntary Program Guide* (FAA Document AFS-900-002-G201). Washington, DC: Author.

FAA. (2014b). *Safety Management: International Collaboration*. Retrieved from https://www.faa.gov/about/initiatives/sms/international/.

FAA. (2015). *Safety Management System: SMS Pilot Projects Overview*. Retrieved from https://www.faa.gov/about/initiatives/sms/pilot_projects/overview/.

FAA Air Traffic Organization. (2014). *International Strategic Plan, Fiscal Year 2014*. Retrieved from https://www.faa.gov/about/office_org/headquarters_offices/ato/service_units/systemops/ato_intl/media/pdf/2014_ATO_Int_Strategic_Plan_core.pdf.

Franck, R. E., Lewis, I., & Udis, B. (2012). *Impact of Export Controls and Technology Transfer Regimes: International Perspectives*. Retrieved from http://oai.dtic.mil/oai/oai?verb=getRecord&metadataPrefix=html&identifier =ADA568332.

Friend, M. A. (2011). Top management support. In A. Stolzer, C. Halford, & J. Goglia (Eds.), *Implementing Safety Management Systems in Aviation* (pp. 161–267). Burlington, VT: Ashgate.

Friend, M. A., & Kohn, J. P. (2010). *Fundamentals of Occupational Safety and Health* (5th ed.). Lanham, MD: Government Institutes.

Gerber, M., & von Solms, R. (2001). From risk analysis to security requirements. *Computers and Security, 20*(7), 577–584.

Government Accountability Office. (2011). *Next Generation Air Transportation: Collaborative Efforts with European Union Generally Mirror Effective Practices, but Near-Term Challenges Could Delay Implementation* (Report No. GAO-12-48). Washington, DC: Government Publishing Office.

Honeywell. (2015). *Honeywell Forecasts Steady Global Helicopter Demand for Next Five Years*. Retrieved from https://aerospace.honeywell.com/en/about/media-resources/newsroom/honeywell-forecasts-steady-global-helicopter-demand-for-next-five-years.

International Civil Aviation Organization (ICAO). (2006). *Safety Management Manual* (SMM) (1st ed.). ICAO Order No. 9859 AN/474. Quebec, Montréal, Canada: Author.

International Civil Aviation Organization (ICAO). (2013). *Safety Management Manual* (SMM) (3rd ed.). ICAO Order No. 9859 AN/474. Quebec, Montréal, Canada: Author.

International Standards Organization. (2012). *Quality Management Principles*. Retrieved from http://www.iso.org/iso/qmp_2012.pdf.

Klein, K. J., & Knight, A. P. (2005). Innovation implementation overcoming the challenge. *Current Directions in Psychological Science, 14*(5), 243–246.

PricewaterhouseCoopers. (2013). *Aviation's Second Golden Age: Can the US Aircraft Industry Maintain Leadership?* Retrieved from http://www.pwc.com/en_US/us/industrial-products/publications/assets/pwc-commercial-aircraft-industry-future-report.pdf.

Repenning, N. P., & Sterman, J. D. (2002). Capability traps and self-confirming attribution errors in the dynamics of process improvement. *Administrative Science Quarterly, 47*, 265–295.

Research and Markets. (2014). *Strategy Guide – 2015 – World's 6 Leading Helicopter & Rotorcraft Manufacturers – Key Strategies and Plans – Airbus Helicopters, AgustaWestland, Bell Helicopter, Boeing, Russian Helicopters, Sikorsky*. Retrieved from http://www.researchandmarkets.com/reports/3023560/strategy-guide-2015-worlds-6-leading#relb3.

Stolzer, A., Halford, C., & Goglia, J. (2008). *Safety Management Systems in Aviation.* Burlington, VT: Ashgate.

Stolzer, A., Halford, C., & Goglia, J. (2011). *Implementation of Safety Management Systems in Aviation.* Burlington, VT: Ashgate.

Tang, R., & Elias, B. (2012). *Off Shoring of Airline Maintenance: Implications for Domestic Jobs and Aviation Safety* (Congressional Research Service Document R42876). Washington, DC: Government Publishing Office.

Transport Canada. (2001). *Introduction to Safety Management Systems* (Report TP 13739). Retrieved from http://tc.gc.ca/media/documents/ca-standards/1191476.pdf.

Tsiakis, T., & Stephanides, G. (2005). The economic approach of information security. *Computers and Security, 24*, 105–108.

Turner, A. (2013, November 17). Middle East debates central ATM agency. *Aviation International News.* Retrieved from http://www.ainonline.com/aviation-news/dubai-air-show/2013-11-17/middle-east-debates-central-atm-agency.

Yantiss, B. (2011). SMS implementation. In A. Stolzer, C. Halford, & J. Goglia (Eds.), *Implementing Safety Management Systems in Aviation* (pp. 161–267). Burlington, VT: Ashgate.

Appendix A

Example SMS Outline for a Small Operation

COMMITMENT TO SAFETY STATEMENT

1.0 Safety policy and objectives
1.1 Safety accountability
1.2 Appointment of key safety groups and personnel
 1.2.1 Initial Review Committee (IRC)
 1.2.2 Safety Review Board (SRB)
 1.2.3 UAS Oversight Committee (UASOC)
 1.2.4 Accident Investigation Board (AIB)
1.3 Emergency response planning
1.4 UAS operational safety
 1.4.1 Hazard safety zones
1.5 General operational guidance
1.6 Obtaining approval to operate UAS
1.7 Obtaining a Certificate of Authorization/Waiver (COA)
1.8 UAS registration requirements
1.9 Model aircraft/hobbyist use
2.0 Safety risk management
2.1 Risk analysis process
 2.1.2 Hazard identification
 2.1.3 Risk analysis
 2.1.4 Required definitions
 2.1.5 Risk assessment
2.2 Risk control
2.3 Risk assessment and mitigation
 2.3.1 Risk evaluation
 2.3.2 Recording the risk management process

3.0 Safety assurance
3.1 Safety performance monitoring and measurement
    3.1.1 Audits
    3.1.2 External audits
    3.1.3 Self-audits
    3.1.4 Advisory committee audits
    3.1.5 Insurance audits
3.2 Continuous improvement of the SMS
    3.2.1 Safety Management System (SMS) annual review
3.3 Monitoring, review and continuous improvement
3.4 Mandatory incident/accident reporting system19
3.5 Anonymous UAS Safety Reporting System (AUSRS)
3.6 Safety reviews
4.0 Safety promotion
4.1 Training
4.2 Active safety communication

Appendix B

Investigation Management System Events

The following list is intended to serve as guidance for the accident investigator to effectively manage an investigation from beginning to end (OSHA, 2015).

1. Initial response
2. Initial actions at the site
3. Secure flight operations documents
4. Human remains recovery
5. Eyewitness interviews
6. Flight recorder recovery
7. Secure weather documents
8. Secure air traffic services and airport documents
9. Search and rescue operations
10. Secure pertinent cabin documents
11. Secure maintenance documents
12. Examination of systems
13. Examination of structures
14. Examination of engine(s) and propeller(s)
15. Initial survey of the accident site
16. Site photography (Phase 1)
17. Review of operations documents
18. Crew member medical examinations
19. Plot flight path
20. Read-out of flight recorders
21. Review of weather documents
22. Review air traffic services and airport documents
23. Evacuation operations
24. Review pertinent cabin documents

25. Review of maintenance documents
26. Examination and testing (systems)
27. Fire and explosion
28. Examination and testing (powerplants)
29. Wreckage distribution plotting
30. Site photography (Phase 2)
31. Flight crew members interviews
32. Victim identification
33. Interviews of next of kin
34. Analysis of flight recorders data
35. Interviews (meteorology)
36. Interviews (air traffic services and airport)
37. Rescue operations
38. Cabin condition
39. Interviews (maintenance and records)
40. Interviews (systems)
41. Crashworthiness
42. Aircraft performance
43. Autopsies
44. Re-interviews (eyewitnesses)
45. Navigation aids and airport status
46. Firefighting operations
47. Interviews (cabin crew and passengers)
48. Maintenance management
49. Wreckage reconstruction
50. Analysis and report of operations group
51. Analysis and report of medical/human factors group
52. Analysis and report of witness group
53. Analysis and report of flight recorders group
54. Analysis and report of meteorology group
55. Analysis and report of ATS and airport group
56. Analysis and report of survivability group
57. Analysis and report of cabin safety group
58. Analysis and report of maintenance and records group
59. Analysis and report of systems group
60. Analysis and report of structures group
61. Analysis and report of powerplants group
62. Analysis and report of site survey group
63. Analysis and report of photo/video group
64. Operations analysis and findings
65. Technical analysis and findings
66. Report of the investigator-in-charge

Appendix C

Investigation Event Task-Assignment Chart

Investigation Group	Assigned Events								
Administration Support									
Head Office Coordinator									
Media Coordinator									
Site Safety									
Investigator-in-charge									
Deputy Investigator-in-charge									
Operations									
Aircraft Performance									
Human Factors									
Medical and Pathology									
Witness									
Flight Recorders									
Meteorology									
Air Traffic Services									
Airports									
Survivability									
Cabin Safety									
Maintenance and Aircraft Records									
Systems									
Structures									
Crashworthiness									
Powerplants									
Site Survey									
Photo/Video									

Figure A.C All activities of the investigation group should be well documented for ease of monitoring using an investigation event-task assignment chart.

Index

Page references for figures are italicized.

About the Editors and Contributors

ABOUT THE EDITORS

Mark A. Friend, EdD, CSP

Dr. Mark Friend, CSP, serves as a program coordinator for the master of science in Occupational Safety Management and professor in the School of Graduate Studies in the College of Aviation at Embry-Riddle University (ERAU) in Daytona Beach, Florida. He has been active as a consultant, speaker, expert witness, and trainer in the field of safety. His text, *Fundamentals of Occupational Safety and Health,* a top-selling book in the field, is in its seventh edition. He is also co-editor of *Planning and Managing the Safety System* with the late Ted Ferry. Friend served on the governing board of the Board of Certified Safety Professionals (BCSP) and is a professional member of the American Society of Safety Professionals (ASSP). He served on the editorial board of *Occupational Hazards* and has published several articles in safety and safety management. His work in grants and contracts includes funding from the National Institute for Occupational Safety and Health (NIOSH), the National Institute of Standards and Technology (NIST), the United States Coast Guard, the Federal Aviation Administration (FAA), and the United States Trade and Development Agency (USTDA) through the Chinese Aviation Cooperation Program (ACP), the Akwa Ibom State of Nigeria, and numerous other public and private entities.

Alan J. Stolzer, PhD, FRAeS

Dr. Alan Stolzer, FRAeS has more than thirty years of experience in academic administration positions and is currently the dean of the College of Aviation and professor of doctoral studies, Embry-Riddle Aeronautical University (ERAU), Daytona Beach Campus. Stolzer is an active researcher and was the principal investigator on several recent federally funded grants related to aviation safety. He is the lead author on two books on safety management systems (SMS). Stolzer's research centers on the subjects of SMS, aviation safety information analysis and sharing systems, and flight operations quality assurance. Stolzer holds a PhD in quality systems from Indiana State University and holds several professional certifications: Quality Engineer, Quality Manager, Quality Auditor from the American Society for Quality; Project Management Professional from the Project Management Institute; Airline Transport Pilot, Flight Instructor with Instrument and Multi-Engine Ratings, and Airframe and Powerplant Mechanic from the Federal Aviation Administration.

Marisa D. Aguiar, MS

Marisa D. Aguiar is a PhD in aviation candidate at Embry-Riddle Aeronautical University working toward completing her dissertation in aviation safety. Throughout the pursuit of her PhD, Marisa has held the position of doctoral graduate research assistant within the College of Aviation's School of Graduate Studies and has had the opportunity to serve as both lead author and coauthor on numerous publications in diverse areas of aviation. Post-graduation, Marisa plans to enter a career in academia and continue to actively pursue her research interests.

ABOUT THE CONTRIBUTORS

Abraham Adole

Captain Abraham Adole is a helicopter pilot in the Nigerian Air Force. His academic achievements include a bachelor of science, a post-graduate diploma in management, and a master's in national security and war studies. Abraham is currently pursuing his PhD in Aviation from Embry-Riddle Aeronautical University.

Susan K. Archer, EdD

Dr. Archer is an adjunct instructor in the Department of Mathematics for Embry-Riddle Aeronautical University's Worldwide Campus. Dr. Archer obtained her doctor of education in educational leadership from the University of North Florida.

Bradley S. Baugh

Brad is a PhD in aviation candidate at Embry-Riddle Aeronautical University. As a doctoral graduate research assistant, he has authored and coauthored numerous journal articles largely in the areas of human factors in aviation and unmanned systems. Prior to retiring from the US Air Force, Brad served as an Airfield Operations Officer specializing in airfield management, air traffic control, emergency preparedness, and safety program management.

David A. Carroll
David Carroll is the Deputy Senior Upset Prevention and recovery training instructor for Aviation Performance Solutions LLC providing instructor quality assurance and mentorship to the US Army's fixed-wing aviation training programs. After retiring from the US Air Force with twenty-four years of service, David has experience at the tactical, operational, and strategic levels of management. He has about 6,000 flight hours and has provided over 2,300 hours of flight instruction. David is currently pursuing his PhD in aviation from Embry-Riddle Aeronautical University.

Stacie L. Fain
Stacie Fain is an airport manager for Steamboat Springs Airport (KSBS) and a captain in the United States Coast Guard Reserve. Stacie is an accomplished program manager, business development professional, and subject matter expert with over twenty-five years of aviation, maritime, military, homeland security, and defense experience supporting multiple U.S. Coast Guard and Department of Defense programs. She is currently pursuing his PhD in aviation from Embry-Riddle Aeronautical University.

Michael Goldfein
Michael "Grip" Goldfein is a captain at Southwest Airlines. Grip received his bachelor of science in basic academics from the United States Air Force Academy. He also holds a master's in aeronautical science degree and is actively pursuing his PhD in aviation from Embry-Riddle Aeronautical University.

Christopher Groom
Christopher Groom is currently serving in Guam as a member of the US Air Force. He has fourteen years of experience working in integrated avionics as an aircraft specialist and presently leads a maintenance production team as an element leader. Christopher holds a bachelor of science in operations management, a master of business specializing in aviation, and is actively pursuing his PhD in aviation from Embry-Riddle Aeronautical University.

Roger C. Lee

Roger Lee is currently the acting director of Safety, Health, Environment, and Quality for Power Systems at CLP Utilities in Hong Kong. Roger holds a master of engineering from Cambridge University and has qualified as a UK Engineering Council Charted Engineer, a fellow of Hong Kong Institution of engineers, air safety investigator, and lead auditor. Roger's aviation experience spans from avionic systems design and integration, Air Traffic Management Systems, Project Management to Safety, Security and Environmental Management for Air Traffic Management Organizations, government aviation agencies/regulators, airports, business aviation and commercial aviation (airlines) operators. Roger is currently pursuing his PhD in aviation from Embry-Riddle Aeronautical University.

Isaac Munene, PhD

Dr. Isaac Munene is currently an aircraft system safety engineer at Bell Flight. He is responsible for aircraft level safety assessments including Aircraft Functional Hazard Assessment, Preliminary Aircraft Safety Assessment, Aircraft Safety Assessment, Zonal Safety Assessment, and Particular Risk Analyses. Dr. Munene holds a bachelor of science in aerospace engineering, a master of aeronautical science dual specializing in Aviation Safety and Management, and a PhD in aviation specializing in Aviation Safety and Human Factors.

Kris Ostrowski, PhD

Lt Col Kris Ostrowski, PhD is an aerospace and operational physiologist for the US Air Force. Kris is also a flight safety officer and former deployed chief of safety specializing in remotely piloted aircraft and traditional aircraft mishap analysis. As a human systems integration consultant, he has participated in various aeronautical projects to include aircraft design and development, test and evaluation, and retrofit. Dr. Ostrowski received his PhD in aviation specializing in safety and human factors from Embry-Riddle Aeronautical University in 2016.

Mark Simpson

Mark Simpson has spent the last few years working for Textron Systems focusing on small, medium endurance unmanned aerial systems (UAS). He has served as a flight instructor, maintainer, and check pilot. He has also taught UAS courses at a number of universities. He currently oversees a team of courseware developers for numerous platforms and products for Textron Systems. Mark has a bachelor of science in aeronautics with a focus on UAS systems, a master of science in geospatial intelligence, and is currently pursuing his PhD in aviation.

Rachelle L. Strong
Rachelle Strong works as a production engineering manager for Boeing Commercial Airplanes and is currently focusing on fuselage manufacturing, equipment, and tool engineering for the Boeing 777 and 777X. In addition to being a certified flight instructor and holding a Commercial Pilot Single and Multi-Engine Certification, Rachelle is currently pursuing a PhD in aviation specializing in human factors and aviation safety with specific research emphasis on pilot fatigue and fatigue monitoring technologies.

Sarah Talley
Sarah Talley is a site program manager for Lockheed Martin at Hill Air Force Base. With expertise in Supply Chain and Manufacturing, Sarah has extensive experience in the realm of project management. Sarah has received a bachelor in aeronautical science, a master of science in aeronautics, and is currently pursuing her PhD in aviation from Embry-Riddle Aeronautical University.

David Thirtyacre
David Thirtyacre is an assistant professor in the College of Aeronautics at Embry-Riddle Aeronautical University's Worldwide Campus and is serving as the chair of the Worldwide Department of Flight. David is actively pursuing a PhD in aviation from Embry-Riddle Aeronautical University. Prior to entering academia, David retired from the US Air Force with twenty-seven years of duty in the fighter community with over 3,500 hours in fighter aircraft and was a pioneer in the use of unmanned systems.

Cory J. Trunkhill
Cory Trunkhill is currently a program manager for Aerotek. His experience includes leading diverse vocations into singular continuity, managed high-value assets, and supervised complex operations related to space and missile warning, command and control nodes, and strategic radar assets. Cory is also currently pursuing his PhD in aviation from Embry-Riddle Aeronautical University.

Gregory S. Woo, PhD
Dr. Gregory S. Woo graduated from Embry-Riddle Aeronautical University in 2017 with a PhD in aviation. He is currently the chief of the Aviation Systems Engineering Division at the United States Department of Transportation Volpe Center. Dr. Woo has more than thirty years of broad experience in the areas of data center management, computer systems engineering, project management and consulting, and management of safety risk analysis for both operational processes and systems engineering processes.

www.ingramcontent.com/pod-product-compliance
Lightning Source LLC
LaVergne TN
LVHW090904040326
833019LV00031B/624